MW01439036

"DOES PROLONGED EXPOSURE OF ALBINO MICE TO MEMORY ENHANCING DRUG CAUSE PERTURBATIONS IN NEUROTRANSMITTER SYSTEM? – A REALISTIC APPROACH"

K. SAILAJA, M.Sc.,

ACKNOWLEDGEMENTS

It is my privilege to express my sincere thanks to my Research Supervisor and also the Head **Prof. K. Yellamma,** Department of Zoology, Sri Venkateswara University, Tirupati for suggesting the research topic and for taking special interest in paving my research career and for providing the necessary facilities in the Department. I also thank Prof.W. Rajendra and Prof. K. Sathyavelu Reddy for their encouragement and support.

I express my indebtedness to my father K.P.M.P. Kumar, my mother N. Durga Sita Laxmi and my brothers K. Durga Prasad and K. Pavan Kumar for their constant encouragement throughout this work. I also thank Mr. K. Anjaneyula Raju for caring the animals.

Finally, I extend my sincere thanks to my fellow researchers Mrs. B. Nirmala Kumari, Miss. K. Kalyani Bai, Miss. P. Glory and Miss. K. Aswini for their timely help and encouragement.

KALIDASU SAILAJA

CONTENTS

PREFACE

In developed countries neurological disorders are most suffering problems. Among these the 1st neurological disorder was Alzheimer's Disease. It affects 17–25 million people worldwide, with an estimated four million in the US and an estimated 8,00,000 people in the U.K. The prevalence of dementia in people over the age of 65 is 5% and in people over 80, it is 20%. At present there is no blood test or radiological test to diagnosis Alzheimer's Disease. It is incurable. Various strategies have been investigated to cure AD, but the use of cholinesterase inhibitor drugs has been the most clinically successful. Scientists developed so many drugs which are helpful to increase memory, learning and intelligence because the major symptoms of Alzheimer's Disease were memory loss. The neurotransmitters like ACh, Dopamine, Norepinephrine, Serotonin etc., play major role in learning and memory so the drugs which are developed for increasing memory are mainly targeted to alter these neurotransmitters levels. Now-a-days, in the market so many memory enhancing drinks such as brain Speed Shake, brain Speed Smoothie, Mocha Focus Delight etc., and foods and drugs (nootropics) are available and these memory enhancing drinks are having structural similarities to anti alzheimer's drugs. People are very much fascinated to take these memory enhancing products to boost up their memory. This has given an idea to take up the present study.

In the present investigation, **galantamine hydrobromide,** one of the recently invented and mostly used memory enhancing drug has been chosen to study its prolonged effects on some selected physiological and biochemical parameters in the brain of albino mice. The first chapter deals with the determination of ED_{50} and also behavioural changes in mice on treatment with galantamine hydrobromide.

In the second chapter, the effects of galantamine hydrobromide on different brain areas of mice have been studied on the levels of monoamines and their allied enzyme systems. In the third chapter, changes in ATPase systems were compiled to see the interaction of galantamine hydrobromide with the energy synthesizing calibre of mice brain. In the fourth chapter, the Lipinski's rule of five to some selected drugs used to treat Alzheimer's disease was tested in order to know whether they obey all the conditions of Lipinski's rule of five.

The present study by itself is not an exhaustive one, as it only throws light on the possible effects of galantamine hydrobromide on all neurotransmitters such as i.e., adrenergic, noradrenergic, serotonergic and dopaminergic in different areas of mice brain which resulted in the overall perceptible changes in the behaviour of the animal. The study also highlights the inhibitory effects of galantamine hydrobromide on energy synthesis as well as utilization mechanisms of brain. It is also a preliminary work as much has to be done to arrive at more definite conclusions.

From the above analysis it is concluded that the intake of memory enhancing drinks which contain memory enhancing drugs for a short period has positive effect, but prolonged intake causes adverse effects. In view of these observations, it is not advisable to recommend these health drinks or drugs particularly to children.

CHAPTER - I

CHANGES IN MORPHOMETRIC AND BEHAVIOURAL ASPECTS

Behaviour includes the process by which an animal senses the external and internal state of body. Such process will take place inside the nervous system and may not be directly observed but reflected through the behaviour of the animal. When a drug is administered chronically without disease condition some changes occur from the normal behaviour and they can be observed externally. These are also called as side-effects. These behavioural changes would be caused by the changes in the nervous system directly or through metabolic or physiological activities.

Neurotransmitters are important for memory, learning, and for overall behaviour of animal. Such neurotransmitters include: **Acetylcholine**, which is involved in muscle contractions; dopamine, which is involved in reward or reinforcement, cognition and diseases such as Parkinson's disease, mood disorders, and schizophrenia; **Norepinephrine**, which helps to regulate arousal and moods, excites gastrointestinal activity, and modulates endocrine function (e.g., insulin secretion); Epinephrine or Adrenalin, which is involved in vasoconstriction and dilation, relaxation of smooth muscles of the intestine (thus inhibiting intestinal motility), and endocrine function; **Serotonin** (5-HT), which is found in greatest concentration in the gastrointestinal tract and is involved in sensory perception, mood control, depression, impulsivity, aggression, and other behaviour problems. The amino acids **GABA** and **Glutamate** also act as neurotransmitters. GABA is the main inhibitory neurotransmitter reducing anxiety; glutamate is the main excitatory neurotransmitter and is involved in memory formation.

The present selected drug galantamine hydrobromide is recommended to improve the cognitive functions and subsequently to treat the Alzheimer's Disease patients. But the usage of galantamine hydrobromide in the absence of disease for prolonged period causes some behavioural disturbances after some period of time. This was proved in this chapter by conducting the water maze experiment and observation of some behavioural characters.

Before starting the experiment the animals were acclimatize to maze environment. For the present study, the animals were divided into 48 batches each batch consisting of 6 animals. Among them, 24 batches were labelled as control and remaining 24 batches as experimental. The water maze experiment was conducted for both control and experimental animals on the selected days viz. 15th, 30th, 45th, 60th, 75th, 90th, 105th, 120th, 135th, 150th, 165th, 180th, 195th, 210th, 225th, 240th, 255th, 270th, 285th, 300th, 315th, 330th,

345th and 360th for all six animals in a group separately and the time taken to reach the hidden platform was noted down and average was calculated. By the comparing the time for experimental and control groups, improvement in learning and memory activity was assessed.

Results

Morphometric Studies (Table 2 & Fig. 1)

From the results on morphometric studies, it was observed that in general all the experimental mice gained relatively more body weight from 15th day to 240th day compared to their corresponding controls.On 15th day there was 20% increase in experimental mice compared to control. From 15th day onwards up to 210th day, the experimental mice recorded further gain of 35% in their body weights. However, on 240th day, the body weights of control and experimental mice were same. After 240th day the experimental mice started losing their body weights gradually up to 360th day.

Behavioural Changes (Table 3 & Fig. 2)

The results demonstrated that intake of galantamine caused significant changes in the behavioural aspects of mice. The results on water maze experiment revealed that the experimental animals treated with galantamine hydrobromide were very active compared to control animals. From the results, it was observed that at any given time the experimental animals took less time compared to control animals to reach the hidden platform. On 15th day, control animals took 180 seconds but experimental only 150 seconds. One significant observation was that the control mice on 180th day took <u>127</u> seconds where as the experimental mice only <u>15</u> seconds. Thus there is a nine fold increase in the learning and memory activity of the experimental mice. Another interesting observation was that on 300th day, the time taken to reach hidden platform for both control and experimental animals was almost the same but after 300th day the time was increased for experimental animals compared to control up to 360th day.

In addition to these, the mice developed a number of side-effects like vomiting, anxiety, dizziness, indigestion, weight loss, sleeplessness, urinary tract infection, tremors and convulsions etc., on continuous exposure of galantamine hydrobromide.

Discussion

The results of the present study clearly state that galantamine hydrobromide causes significant changes in the body weight and behavioural aspects of mice. Compared to control the experimental mice gained body weight upto 240 days but after 240 days they started losing their body weight gradually. The significant observation on behavioural aspect was the experimental animals were more active on 180th day.

Animal behaviour is neurally regulated phenomenon mediated by the brain and neurotransmitters **(Bullock *et al.*, 1977)**. Normally cholinesterase inhibitors appear to differ and improve apathy, depression and anxiety. Support to the present study there was also evidence that the rats administered galantamine (2.5mg/Kg/day I.P) showed an improved speed of learning and short-term memory in the shuttle box test **(Iliev *et al.*, 1998)**. It is also reported that AChE inhibitors have been less promising therapeutically as they produce only modest improvements in cognitive function **(Bryson and Benfield, 1997; Green berg *et al.*, 2000; Mohs *et al.*, 2001; Winblad et al., 2001)** along with considerable side-effects **(Wilkinsen, 1999; Dunn *et al.*, 2000; Imbimbo, 2001)**. In addition, studies have shown that the drugs increases the release of other neurotransmitters beside ACh, including Dopamine and Norepinephrine from rat brain regions known to play an important role in learning and memory functions **(Menzaghi *et al.*, 1998)**. In the present study also the selected drug showed increased levels of Dopamine and Norepinephrine upto 180 days after that the levels gradually decreases. In water maze experiment also the learning and memory performance of animals was very well up to 180 days and after it gradually reaches normal.

The pharmacologists observed in the time of evaluating the efficacy of donepezil, rivastigmine and galantamine in control trials they showed significant improvement in cognitive functions up to six months after that they were become normal **(Rockville *et al.*, 1989)**. There was also report which supported to the present study the drugs used to treat AD are having the ability to improve all neurotransmitters systems and causes activeness and learning and memory improvements in disease or in without disease condition for only some period and after that the remaining normal **(Molloy *et al.*, 1991)**. There was also evidence that the long-term efficacy of galantamine is delaying the cognitive function, activities of daily living and behavioural disturbances has recently began to be reported **(Pirltila *et al.*, 2004 and Chun *et al.*, 2007)**.

The central noradrenergic system is considered to play important roles in attention, learning and memory processes (**Aston-Jones *et al.*, 1991; Ferry *et al.*, 1999a, b, Berridge *et al.*, 1993; Sara et al., 1994**). Noradrenaline has also been found to have an important permissive role in long-term potentiation, a form of synaptic plasticity associated with memory processes (**Bliss *et al.*, 1983**) and also facilitates in a synergistic manner a similar role for ACh in long-term potentiation (**Brocher *et al.*, 1992**). Decline in memory is one of the major symptoms of Alzheimer's Disease, and this impairment has been ascribed to the central cholinergic (**Perry *et al.*, 1992**) and noradrenergic (**Reinkainen *et al.*, 1990; Haapalinna *et al.*, 1998**) pathologies that occur in this disorder. The memory enhancing effects of drugs could also in part to the result of receptor antagonists induced facilitation in the release of neurotransmitters such as Dopamine or Serotonin (5-HT) (**Raiteri *et al.*, 1990**) which also modulate learning and memory processes.

The present study reveals that the uptake of galantamine hydrobromide in the absence of disease condition improves neurotransmitter release such as Dopamine, Norepinephrine etc., and also thereby improves learning and memory but only 300 days, after that it shows side effects like weight loss, vomiting, tiredness, dizziness etc. For evidence to this there was also reports that the memory enhancing drugs like opiates and other drugs taking for a long time causes anxiety like effects and other side effects (**Shaham *et al.*, 2003; Lu *et al.*, 2003**). There was also reports that the effects of memory enhancing drugs like galantamine hydrobromide are similar to cigarette smoking. Cigarette smoking temporarily normalizes sensitomotor gating deficits in schrizophrenics but prolonged usage causes so many health problems (**Adler *et al.*, 1993; Kumari *et al.*, 2001**). The other reports reveals that chronic administration of nicotine or any other memory enhancing drugs (galantamine, haloperidol, risperielone and clozapine) reduced locomotor activities in animals and attentional improvement (**Rezvani and Levin, 2004; Rezvani *et al.*, 2006**).

An interesting observation in the present study was that the morphometric and behavioural changes in mice treated with galantamine hydrobromide showed positive result up to around 240 days and then onwards, up to 360 days the experimental animals lost their body weights and activity levels because they develop side effects. So it may be suggested that the effect of the drug not only depends on its concentration but on the duration of exposure also. The other factors on which the effect depends include the extent and rate of absorption from the site of administration into the blood stream, distribution to various parts of the body from the blood, binding in tissues and mechanism of inactivation. Finally, it may be concluded that the general ill-health and overall decline in physical ability due to disorders in Central Nervous System in combination pose a great threat on the survival of mice.

Table 2

Differences in the body weight of control and experimental mice at selected time intervals during the treatment with Galantamine hydrobromide for 360 days.

	15d	30d	45d	60d	75d	90d	105d	120d	135d	150d	165d	180d
Control	16	17	18	20	22	22	24	24	23	25	26	28
Experimental	18	22	25	26	28	27	30	31	29	32	30	35
	195d	210d	225d	240d	255d	270d	285d	300d	315d	330d	345d	360d
Control	29	34	36	38	38	41	40	42	41	40	42	43
Experimental	38	45	40	38	35	33	33	31	28	27	25	25

Fig.1: Graphical representation of differences in the body weight of control and experimental mice at selected time intervals during the treatment with Galantamine hydrobromide for 360 days.

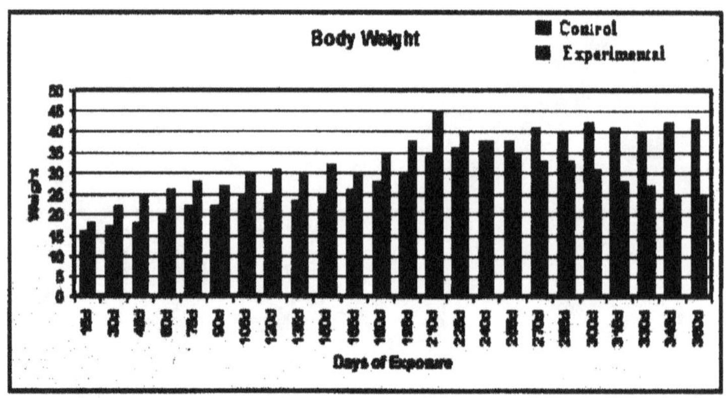

Table 3

The results of the water maze experiments on control and experimental mice treated with Galantamine hydrobromide for 360 days at selected time intervals.

	15d	30d	45d	60d	75d	90d	105d	120d	135d	150d	165d	180d
Control	180s	170s	180s	160s	170s	165s	160s	150s	140s	140s	130s	127s
Experimental	150s	135s	125s	120s	105s	95s	90s	80s	70s	55s	40s	15s

	195d	210d	225d	240d	255d	270d	285d	300d	315d	330d	345d	360d
Control	125s	135s	132s	127s	130s	135s	138s	142s	145s	148s	142s	143s
Experimental	30s	49s	60s	82s	95s	102s	125s	139s	155s	159s	168s	178s

Fig. 2: Graphical representation of the results of the water maze experiments on control and experimental mice treated with Galantamine hydrobromide for 360 days at selected time intervals.

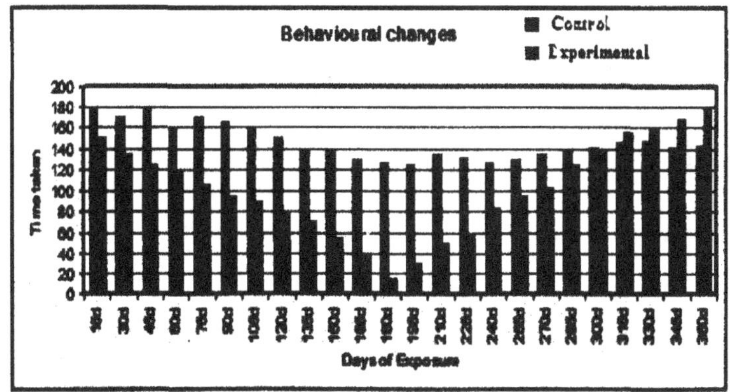

CHAPTER – II

AMINERGIC SYSTEM

Various constituents of aminergic system (**Catecholamines or Biogenic amines**) are sympathomimetic "fight-or-flight" hormones that are released by the adrenal glands in response to stress. They are part of the sympathetic nervous system. They are called catecholamines because they contain a catechol group, and are derived from the amino acid tyrosine (**Purves *et al.*, 2008**). More specifically, catecholamines contain a nucleus of catechol (a benzene ring possessing two adjacent hydroxyl groups) and a side chain of ethylamine or one of its derivatives. The catecholamines belong to the wider group of neurotransmitters called monoamines, that is compounds possessing a single amine ($-NH_2$) group. The most abundant catecholamines are dopamine, norepinephrine (noradrenaline) and epinephrine (adrenaline) all of which are produced from phenylalanine and tyrosine which are aromatic aminoacids derived from dietary protein. Tyrosine is also synthesized in the liver by the enzyme phenylalanine hydroxylase. Tyrosine is then transported to catecholamine secreting neurons where a series of following reactions takes place which leads to the production of catecholamines (**Shiman *et al.*, 1971**). Many catecholamines serve as structural motifs for various stimulants. For example, MDMA (Ecstasy) can be thought of as a reduced product of a double condensation of dopamine with formaldehyde (**Joh and Hawango, 1987**). Epinephrine and norepinephrine are sometimes called adrenaline and noradrenaline respectively. Catecholamines are water - soluble and are 50% bound to plasma proteins, so they circulate in the blood stream. The synthetic pathway for catecholamines was first predicted by **Blaschko (1939)**, who correctly proposed that Dopamine was a precursor for norepinephrine and epinephrine.

Dopamine is the first catecholamine to be synthesized from L-Tyrosine as shown below. Norepinephrine and epinephrine, in turn, are derived from further modifications of dopamine. The rate limiting step in the biosynthesis is hydroxylation of tyrosine. Catecholamine synthesis is inhibited by alpha-methyl-p-tyrosine (AMPT), which inhibits tyrosine hydroxylase (Wrongdiagnosis.com).

Biochemical pathways in Aminergic system

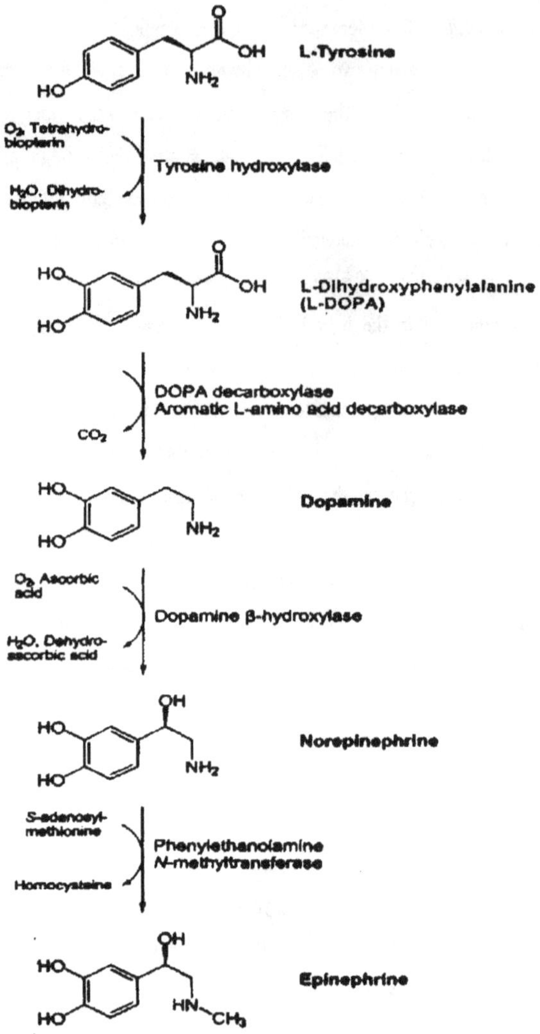

Catecholamines are produced mainly by the chromaffin cells of the adrenal medulla and the postganglionic fibers of the sympathetic nervous system. Dopamine, which acts as a neurotransmitter in the central nervous system, is largely produced in neuronal cell bodies in two areas of the brainstem: the substantia nigra and the ventral tegmental area. Catecholamines are located in various brain regions and other tissues. They serve many functions like emotion, attention and visceral regulation. The catecholamines have distinct pathways in the brain **(Cooper *et al.*, 1982)** which are visualized by anatomical, autoradiography and immunochemical techniques **(Lindvall and Blockland, 1978)**.

Catecholamines have a half-life of a few minutes when circulating in the blood. They can be degraded either by methylation by catechol-*O*-methyltransferases (COMT) or by deamination by monoamine oxidases (MAO). Amphetamines and MAO inhibitors bind to MAO inorder to inhibit its action of breaking down catecholamines. This is the primary reason why the effects of amphetamines have a longer lifespan than those of cocaine and other substances. Amphetamines not only cause a release of dopamine, epinephrine and norepinephrine into the blood stream but also suppress re-absorption.

Two catecholamines viz. norepinephrine and dopamine act as neuromodulators in the central nervous system and as hormones in the blood circulation. The catecholamine norepinephrine is a neuromodulator of the peripheral sympathetic nervous system but is also present in the blood (mostly through "spillover" from the synapses of the sympathetic system). High catecholamine levels in blood are associated with stress, which can be induced from psychological reactions or environmental stressors such as elevated sound levels, intense light or low blood sugar levels. Extremely high levels of catecholamines (also known as catecholamine toxicity) can occur in the central nervous system trauma due to stimulation and/or damage of nuclei in the brainstem, in particular those nuclei affecting the sympathetic nervous system. In emergency medicine, this occurrence is widely known as *catecholamine dump*. Extremely high levels of catecholamine can also be caused by neuroendocrine tumors in the adrenal medulla, a treatable condition known as pheochromocytoma. High levels of catecholamines can also be caused by monoamine oxidase A deficiency which is one of the enzymes responsible for degradation of these neurotransmitters and thus increases the bioavailability of them considerably. It occurs in the absence of pheochromocytoma, neuroendocrine tumors, and carcinoid syndrome, but it looks similar to carcinoid syndrome such as facial flushing and aggression **(Manor *et al.*, 2002).**

The most commonly observed actions of catecholamines are depressant, seen in the lateral geniculate **(Phillis and Tebecis 1967a)**, hippocampus **(Biscoe and Straughan, 1966)**, pyriform cortex **(Legge *et al.*, 1966)**, hypothalamus **(Bloom *et al.*, 1963)**, olfactory bulb **(Salmoiraghi *et al.*, 1964)**, straitum **(Bloom et al., 1965)**, thalamus **(Phillis and Tebecis, 1967a)**, medial geniculate (Tebecis, 1970), spinal cord **(Engberg and Ryall, 1966)**, red nucleus **(Davis and Vaughan, 1969)**, medulla **(Boakes *et al.*, 1972)**, cerebellum **(Siggins *et al.*, 1971)**, midbrain reticular formation and superior colliculus **(Straschill and Perwein, 1971)**. Catecholamines cause general physiological changes that prepare the body for physical activity (fight-or-flight response). Some typical effects are increase in heart rate, blood pressure, blood glucose levels and a general reaction of the sympathetic nervous system. Some drugs, like tolcapone (a central COMT-inhibitor), raise the levels of all the catecholamines.

Dopamine

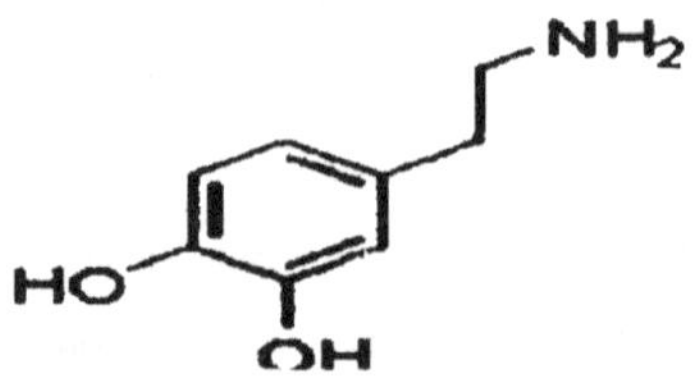

NH_2

HO OH

Several investigators have shown that learning and memory can be modified by drugs which affect the central dopamine (DA) neuronal system **(Barret *et al.*, 1974).** Dopamine represents 50% of the total catecholamines content in the central nervous system. The highest levels of dopamine are found in the neostriatum, nucleus accumbans and tuberculum olfactorium. Dopamine synthesis was dependant on the aminoacid precursor or tyrosine, which must be transported across the blood-brain- barrier into the dopaminergic neurons. Dopamine also acts as an important neurotransmitter in the peripheral nervous system, affecting both gastrointestinal and cardiovascular activity **(Bell, 1987).** Dopamine levels were high in the nucleus basalis and very low levels were found in optic lobes, cerebellum and spinal cord of pigeon brain **(Jurio and Vogt., 1967).** Dopaminergic lesioning of the prefrontal cortex leads to cognitive dysfunctions and dopamine depletion of the mesoseptal dopaminergic projections leads to a decrease in working memory **(Simon *et al.*, 1986)** and lesioning of the meso accumbans dopaminergic projections cause attention deficit and alterations in locomotor activity **(Carey and Schwarting, 1986).** These findings give insight into dopamine's tardive dyskinesia. Dopamine is involved in reinforcement, generation of pleasure development of drug addiction and so forth **(Fallon and Loughlin, 1995).** Dopamine is the primary neuroendocrine inhibitor of the secretion of prolactin from the anterior pituitary gland **(Ben-Jonathan and Hnasko, 2001).** Dopamine has been shown to be involved in the control of movements, the signaling of error in prediction of reward, motivation, and cognition. Cerebral dopamine depletion is the hallmark of Parkinson's disease **(Arias-Carrión and Pöppel, 2007).** Other pathological states have also been associated with dopamine dysfunction, such as schizophrenia, autism, and attention deficit hyperactivity disorder as well as drug abuse **(Fallon and Loughlin, 1995).**

Norepinephrine

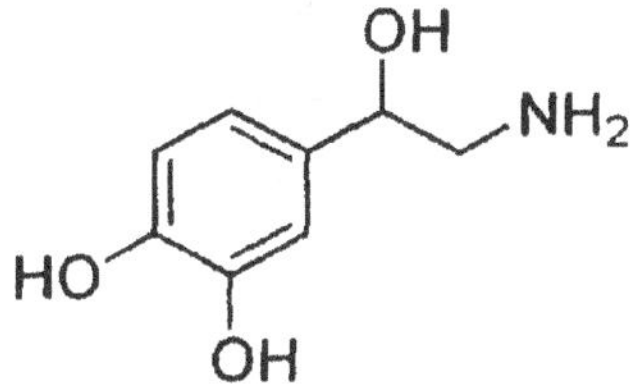

OH
NH_2
HO
OH

Norepinephrine neurons are clustered in midbrain, pons and medulla oblongata and are considered to be a part of reticular formation. On the basis of majority of the axonal projections they can be roughly grouped into two major pathways, they are dorsal and ventral bundles. The noradrenergic axons are scattered in cerebellum the entire cerebral cortex and hippocampus. The ventral bundle is mainly scattered in the pons, medulla and also in hypothalamus **(Snyder and Coyle, 1981).** When a neuron is working right, it releases norepinephrine through two autoreceptors into the synapse and attaches to $\alpha1$, $\alpha2$ and $\beta1$ receptors on the other side of the synapses **(Rohler & Kobilka, 1998)**. From there a signal is sent into the cell that activates certain genes to switch on proteins governing all activity.

Norepinephrine is believed to play a role in cognition, mood emotions and blood pressure. Since norepinephrine doesn't cross the blood brain barrier, those seeking natural treatment must opt for its building blocks. Tyrosine, a precursor of both norepinephrine and dopamine can act as an energizer to natural treatment advocates **(Siegal *et al.*, 1981)**. Norepinephrine is a well established neurotransmitter in the central nervous system with widespread distribution throughout the brain and has hypothesized functions in arousal, attention, anxiety and affective disorders.

Cell bodies for the brain norepinephrine originate in the dorsal pons and brain stem. The locus coruleus in the dorsal pons is the source of the dorsal noradrenergic pathway to the cortices in the ventral noradrenergic bundle to innervate the basal forebrain and hypothalamus. There is well documented evidence that hypothalamus contains particularly large amounts of norepinephrine **(Ungerstedt, 1971a,b)** that direct applications of norepinephrine in the region of hypothalamus and the norepinephrine released in the hypothalamus **(Stein and Wise, 1969)** lead to changes in body temperature **(Feldberg and Myers, 1964)** or in behaviour **(Fitzsimons, 1972)**. Along with epinephrine, norepinephrine also underlies the fight-or-flight response, directly increasing heart rate, triggering the release of glucose from energy stores, and increasing blood flow to skeletal muscle. Norepinephrine also has a neurotransmitter role when released diffusely in the brain as an antiinflammatory agent **(Heneka *et al.*, 2010)**.

Epinephrine

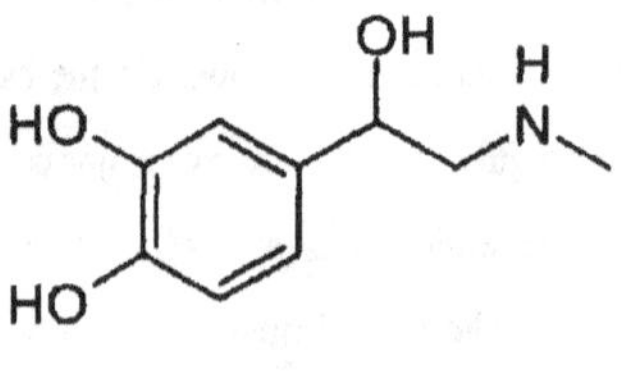

Epinephrine levels are high in hypothalamus in the mammalian brain **(Gunne, 1963)**. Epinephrine concentration in the mammalian central nervous system is relatively low, approximately 5 to 17% of the gray matter than in white matter **(Cooper *et al.*, 1982)**. In addition to their effects as neurotransmitters, norepinephrine and epinephrine can influence the rate of metabolism. This influence works both by modulating endocrine function such as insulin secretion and by increasing the rate of glycogenolysis and fattyacid mobilization.

Serotonin (5-HT)

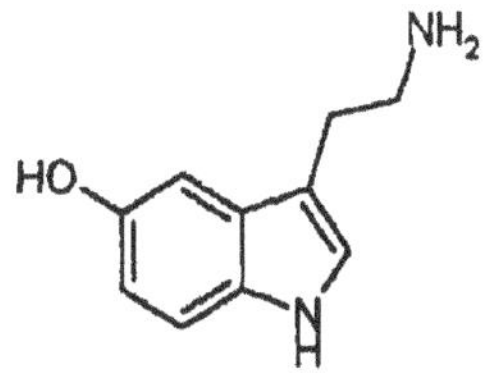

5-HT is synthesized in the neurons from the aminoacid tryptophan, which is converted to 5-HTP and then to serotonin. It is released into synapse in a similar manner to norepinephrine. Serotonin has 17 different types and subtypes of receptors, which underscores its importance as a neurotransmitter. Serotonin fibers project from the raphe nucleus in the brain stem to the basal ganglia, frontal cortex, hypothalamus and limbic system and down the spinal cord. Most of the remainder of the bodies 5-HT is found in platelets of the central nervous system thus implicating itself in a host of functions from mood to anxiety of sleep **(Mc Manamy, 2005)**. Serotonin is a central neurotransmitter that is believed to have an important role in the regulation of mood, sleep, appreciation of pain, appetite behavior, drinking, respiration, heart rate, rhythmic behavior and memory. The hypothalamus, midbrain and brain stem have high concentrations, where as the cerebral cortex, hippocampus and striatum have moderate concentrations and the cerebellum has a low concentration **(Savedra *et al.*, 1973)**. 5-HT plays a role in coordinating specific behaviors including aggression **(Weber *et al.*, 1997)** and feeding **(Lin *et al.*, 2000)**.

The neurotransmitter serotonin is involved in the regulation of basic physiological functions such as hormone secretion **(Hanlay and Van de kar, 2003)**, sleep-wake cycle **(Ursin, 2002)**, motor control **(Jacobs and Fornal, 1997)**, immune system functioning **(Mossner and Lesch, 1998)**, nocireption **(Eide and Hole, 1993)**, food intake **(Meguid *et al.*, 2000)** and energy balance **(Heisler *et al.*, 2003)**. In addition, 5-HT participates to higher brain functions, such as cognition and emotional states, by modulating synaptic plasticity **(Foehring and Lorenson, 1999)** and as recently discovered, neurogenesis **(Djavadian, 2004)**. 5-HT is one of the neurotransmitters used by a descending system controlling pain transmission.

Monoamine Oxidase

MAO is an enzyme that plays an important role in monoamine neurotransmitter metabolism **(Petrovic *et al.*, 2001)**. It is present in invertebrates **(Sloley, 1994)** and vertebrates **(Petrovic *et al.*, 2001)**. It is a flavoprotein enzyme located in the mitochondrial fraction of the liver, kidney and brain **(Coasta, 1972)**. It catalyzes the oxidative deamination of monoamines **(Sote-otero *et al.*, 2001)**.`MAO-A and MAO-B, two isoforms of monoamine oxidase (MAO), are expressed on the mitochondrial outer membrane. MAO mediated neurodegeneration can result from the formation of hydrogen peroxide (H_2O_2) as a by product of metabolism of aminergic neurotransmitters including serotonin and dopamine. If it is not detoxified by antioxidant systems such as glutathione peroxidase – one of the most abundant such systems in brain **(Sies, 1993)** then H_2O_2 can be converted by iron – mediated Fenton reactions to hydroxyl radicals that can initiate lipid peroxidation and cell death. This is exacerbated when antioxidant systems are compromised such as during aging **(Zhu *et al.*, 2006)**. MAO – A also plays a role in neuropsychiatric and behavioral disorders.

RESULTS
DOPAMINE (Tables 4.1 & 4.2 and Figs. 3 to 8)

The results of the present study clearly state that galantamine hydrobromide significantly affected the dopamine levels in all areas of mice brain under prolonged exposure. It was also noted that the effect was more pronounced on 180th day.

On 30th day of galantamine hydrobromide exposure, significant elevation in Dopamine content was observed in all brain areas of mice studied and increase was in the following order.

Hc > Cb > OL > Spc > Pm > CC
(39.77%) (36.43%) (30.76%) (30.06%) (29.50%) (25.59%)

It was observed that, on 60th day different regions of mice brain exhibited different percent changes against control and they were in the following manner.

Cb > OL > Hc > Pm > Spc > CC
(48.75%) (47.15%) (41.10%) (37.45%) (34.06%) (31.42%)

Similar to 60th day recordings, on 90th day also Cerebellum exhibited highest elevation and Cerebral Cortex least.

Cb > OL > Spc > Hc > Pm > CC
(51.05%) (50.18%) (45.68%) (44.93%) (41.30%) (38.71%)

On 120th and 150th day of galantamine exposure the dopamine elevation pattern in different regions was as follows.

120th day OL > Cb > Spc > Hc > Pm > CC
(60.36%) (58.37%) (55.75%) (50.86%) (45.10%) (41.65%)

150th day OL > Cb > Spc > CC > Hc > Pm
(66.39%) (66.34%) (61.64%) (55.43%) (55.37%) (48.58%)

On 180th day, all the regions of mice brain showed **maximum elevation** in dopamine content except Spinal cord. In Spinal cord maximum elevation was observed on 195th day.

OL > Cb > Hc > Spc > CC > Pm
(87.67%) (69.13%) (66.31%) (64.22%) (65.07%) (51.33%)

From 210th day onwards, the dopamine levels showed gradual depletion up to 360th day. For example, on 210th day the percent change in dopamine level against control was recorded in the following order.

OL	>	Spc	>	Cb	>	CC	>	Hc	>	Pm
(62.94%)		(62.02%)		(61.7%)		(59.91%)		(56.79%)		(49.90%)

However, on days 240th and 270th, galantamine hydrobromide effect on dopamine levels was different in different areas and the trend in percent change against control were the same.

240th day	OL	>	Cb	>	Spc	>	Hc	>	Pm	>	CC
	(57.02%)		(54.32%)		(52.66%)		(50.62%)		(44.40%)		(41.53%)
270th day	OL	>	Cb	>	Spc	>	Hc	>	Pm	>	CC
	(52.91%)		(52.65%)		(44.02%)		(43.17%)		(41.16%)		(39.97%)

On days 300th, 330th and 360th the depleting trend in dopamine content of various brain regions continued further in the following order.

300th day	OL	>	Pm	>	Cb	>	CC	>	Hc	>	Spc
	(49.08%)		(37.69%)		(37.15%)		(33.84%)		(33.48%)		(32.26%)
330th day	Cb	>	OL	>	Hc	>	Spc	>	CC	>	Pm
	(37.53%)		(36.36%)		(30.47%)		(27.97%)		(27.15%)		(24.51%)
360th day	Cb	>	CC	>	OL	>	Pm	>	Spc	>	Hc
	(27.27%)		(23.61%)		(22.56%)		(20.72%)		(15.58%)		(12.39%)

NOREPINEPHRINE (Tables 5.1 & 5.2 and Figs. 9 to 14)

As in the case of dopamine, norepinephrine levels were also elevated from 30th day to 180th day following prolonged dosage of mice with galantamine hydrobromide. On days 30th, 60th, 90th and 120th days, it was obvious that norepinephrine levels in different regions of mice brain were almost similar in their trend as mentioned below, but for a small deviation between ponsmedulla and olfactory lobe on 90th day.

30th day	CC	>	Pm	>	OL	>	Hc	>	Cb	>	Spc
	(35.44%)		(34.06%)		(33.02%)		(27.12%)		(20.29%)		(18.25%)
60th day	CC	>	Pm	>	OL	>	Hc	>	Cb	>	Spc
	(44.88%)		(44.40%)		(43.75%)		(33.74%)		(32.61%)		(25.79%)

90th day CC > OL > Pm > Hc > Cb > Spc
(52.30%) (49.06%) (47.10%) (39.62%) (36.44%) (33.71%)

120th day CC > Pm > OL > Hc > Cb > Spc
(60.48%) (57.60%) (53.60%) (51.37%) (45.48%) (37.18%)

On 150th day, a further increase in norepinephrine was observed in different regions of experimental mice brain in the following order.

CC > Hc > Pm > OL > Cb > Spc
(69.40%) (68.14%) (67.40%) (58.56%) (51.21%) (43.61%)

On 180th day, all the regions of mice brain recorded **maximum elevation** in norepinephrine levels except Spinal cord and Cerebral Cortex.

Hc > Pm > CC > OL > Cb > Spc
(78.90%) (74.10%) (73.61%) (63.46%) (61.88%) (48.75%)

From 210th day onwards upto 360th day, the trend started reversing as evidenced by a decrease in the content of norepinephrine levels as given below.

210th day Hc > CC > Pm > OL > Cb > Spc
(68.80%) (68.5%) (61.2%) (57.80%) (56.70%) (44.98%)

240th day Hc > CC > Pm > OL > Cb > Spc
(63.34%) (58.66%) (56.01%) (53.49%) (47%) (39.99%)

270th day Hc > CC > Pm > OL > Cb > Spc
(53.23%) (47.20%) (44.4%) (45.30%) (45.25%) (34.15%)

300th day Hc > OL > CC > Cb > Pm > Spc
(43.71%) (42.98%) (39.48%) (38.86%) (34.4%) (29.92%)

330th day Hc > CC > OL > Pm > Cb > Spc
(33.99%) (33.60%) (32.44%) (29.12%) (23.60%) (22.88%)

360th day OL > Pm > Hc > CC > Cb > Spc
(28.09%) (26.69%) (26.44%) (24.64%) (20.65%) (19.71%)

EPINEPHRINE (Tables 6.1 & 6.2 and Figs. 15 to 20)

As in the case of Dopamine and Norepinephrine, Epinephrine content was also elevated in all brain areas of mice following prolonged dosage with galantamine hydrobromide. On 30th, 60th and 90th days of galantamine hydrobromide exposure the effect was similar in all the regions of mice brain and the trend in percent change against control were same but for a small deviation between Cerebral Cortex and Olfactory Lobe on 90th day.

30th day	Cb	>	Spc	>	CC	>	OL	>	Hc	>	Pm
	(32.4%)		(30.1%)		(25.4%)		(25.3%)		(24.6%)		(22.7%)
60th day	Cb	>	Spc	>	CC	>	OL	>	Hc	>	Pm
	(40.4%)		(34.6%)		(33%)		(31.02%)		(29%)		(27.7%)
90th day	Cb	>	Spc	>	OL	>	CC	>	Hc	>	Pm
	(46.2%)		(43.1%)		(40.3%)		(38.9%)		(35.2%)		(31.1%)

On 120th and 150th day, the effect of galantamine hydrobromide on different brain regions followed a similar trend.

120th day	Spc	>	Cb	>	CC	>	OL	>	Hc	>	Pm
	(54.1%)		(49.8%)		(46.4%)		(45.16%)		(42.1%)		(37.1%)
150th day	Cb	>	Spc	>	CC	>	OL	>	Hc	>	Pm
	(60.4%)		(52.3%)		(51.7%)		(49%)		(46.9%)		(46.2%)

Maximum elevation was observed on 180th day in Olfactory Lobe, Cerebral Cortex and Hippocampus, on 210th day in Cerebellum, on 195th day in Spinal cord and Ponsmedulla. On 180th day, the elevation in epinephrine levels was as follows.

Spc	>	CC	>	Cb	>	OL	>	Pm	>	Hc
(63.1%)		(57%)		(56.8%)		(56.3%)		(51.4%)		(50.5%)

Between 210th to 360th days, a gradual depletion in epinephrine content in different areas of mice brain was observed in the following order.

210th day	Spc	>	Cb	>	OL	>	CC	>	Hc	>	Pm
	(62.3%)		(59.8%)		(53.6%)		(50.8%)		(46.9%)		(44.3%
240th day	Spc	>	OL	>	Cb	>	CC	>	Pm	>	Hc
	(55.23%)		(48.9%)		(48.1%)		(44.5%)		(40.6%)		(40.2%)
270th day	Cb	>	Spc	>	OL	>	CC	>	Hc	>	Pm
	(44.1%)		(42.2%)		(40.5%)		(37.9%)		(33%)		(32.6%)
300th day	Cb	>	Spc	>	CC	>	OL	>	Hc	>	Pm
	(37%)		(36.1%)		(33.7%)		(29.6%)		(27.5%)		(26.7%)
330th day	Cb	>	CC	>	Spc	>	Hc	>	OL	>	Pm
	(28.6%)		(28.5%)		(26.9%)		(22.9%)		(22.8%)		(21.2%)
360th day	Cb	>	Spc	>	CC	>	OL	>	Hc	>	Pm
	(21.1%)		(20.5%)		(18.5%)		(15.7%)		(14.9%)		(14.5%)

Subsequently, it was also observed that on 360th day all the regions in the brain showed almost equal levels of percent change in epinephrine content which was very near to that of 15th day percent changes. From these observations it was evidenced that under the effect of galantamine hydrobromide, epinephrine levels were elevated for some period and after that they again reached normal levels.

SEROTONIN (Tables 7.1 & 7.2 and Figs. 21 to 26)

Fluctuations in Serotonin content were also observed in different regions of mice brain under the prolonged effect of galantamine hydrobromide. On 30th day of galantamine exposure, a significant elevation was observed in all the brain areas studied and increase was in the following order.

Spc >	CC >	Pm >	OL >	Cb >	Hc
(45.27%)	(28.26%)	(25.80%)	(25%)	(23.75%)	(17.10%)

On 60th and 90th days exposure all the regions of mice brain showed a similar trend in the serotonin levels.

60th day	Spc >	Cb >	OL >	CC >	Pm >	Hc
	(49.33%)	(44.33%)	(33.48%)	(31.79%)	(29.58%)	(21.02%)
90th day	Spc >	Cb >	OL >	CC >	Pm >	Hc
	(50%)	(47.7%)	(45.83%)	(37.45%)	(35.42%)	(25.68%)

On 120th, 150th and 180th days the galantamine hydrobromide effect on serotonin levels was different in different areas and the trend on percent change against control were almost the same.

120th day	Spc >	OL >	Cb >	Pm >	CC >	Hc
	(53.42%)	(52.13%)	(50.80%)	(44.46%)	(44.37%)	(37.17%)
150th day	Spc >	OL >	Cb >	Pm >	CC >	Hc
	(68.86%)	(67.43%)	(55.44%)	(52.42%)	(49.61%)	(42.80%)
180th day	Spc >	OL >	Cb >	CC >	Pm >	Hc
	(88.05%)	(71.01%)	(64.45%)	(55.44%)	(54.95%)	(49.42%)

In Spinal cord, Cerebellum and Hippocampus regions **maximum elevation** was observed on 180th day and in remaining regions such as Olfactory Lobe and Cerebral Cortex the maximum elevation was on 165th day and in Ponsmedulla it was on 195th day.

From 210th day onwards the elevated serotonin levels showed a depleting trend upto 300th day and the level of serotonin on 330th day and 360th day were almost near to the 15th day levels.

210th day	Spc >	OL >	Cb >	Pm >	CC >	Hc
	(70.23%)	(63.90%)	(59%)	(57.48%)	(52.69%)	(47.07%)
240th day	Spc >	Cb >	OL >	CC >	Pm >	Hc
	(60.19%)	(54.65%)	(52.55%)	(49.80%)	(47.29%)	(43.44%)
270th day	Spc >	OL >	Cb >	Pm >	CC >	Hc
	(53.42%)	(49.51%)	(48.55%)	(45.10%)	(43.86%)	(41.76%)
300th day	OL >	Pm >	Spc >	CC >	Cb >	Hc
	(42.98%)	(40.71%)	(39.84%)	(36.52%)	(34.59%)	(34.27%)
330th day	OL >	Pm >	CC >	Cb >	Spc >	Hc
	(37.81%)	(29.45%)	(28.75%)	(28.16%)	(26.99%)	(26.47%)
360th day	OL >	Pm >	Spc >	Hc >	CC >	Cb
	(24.63%)	(22.05%)	(21.46%)	(20.88%)	(28.37%)	(17.37%)

MONOAMINE OXIDASE (Tables 8.1 & 8.2 and Figs. 27 to 32)

In contrast with the trend observed in the all of the various constituents of the aminergic system, the MAO enzyme showed a corresponding inhibition in its levels in various regions of mice under prolonged effect of galantamine hydrobromide. From 30th day upto 180th day extent inhibition in MAO in different regions of mice brain were in the following order. However, on 180th day **maximum inhibition** was noticed.

30th day	Spc >	Hc >	CC >	Pm >	OL >	Cb
	(-24.07%)	(-21.05%)	(-18.07%)	(-17.28%)	(-15.07%)	(-15%)
60th day	Spc >	CC >	Hc >	Pm >	Cb >	OL
	(-36.95%)	(-33.42%)	(-22.39%)	(-21.02%)	(-20.57%)	(-19.96%)
90th day	CC >	Spc >	Pm >	OL >	Hc >	Cb
	(-38.83%)	(-38.10%)	(-27.19%)	(-24.57%)	(-24.03%)	(-22.77%)
120th day	CC >	Spc >	Hc >	Pm >	OL >	Cb
	(-47.57%)	(-40.97%)	(-32.98%)	(-30.43%)	(-27.49%)	(-24.70%)
150th day	CC >	Spc >	Cb >	Hc >	Pm >	OL
	(-57.10%)	(-45.90%)	(-43.18%)	(-57.46%)	(-31.88%)	(-31.54%)

180th day	CC >	Spc >	Hc >	Pm >	Cb >	OL
	(-67.01%)	(-49.86%)	(-45.27%)	(-41.19%)	(-41.18%)	(-34.87%)

From 210th day onwards up to 360th day, MAO activity showed gradual increase in all regions of mice brain and on 360th day, the level of MAO was quite nearer to the level noticed on 15th day.

210th day	CC >	Spc >	Hc >	OL >	Pm >	Cb
	(-66.02%)	(-48.46%)	(-41.26%)	(-38.56%)	(-37.07%)	(-32.67%)
240th day	CC >	Spc >	Hc >	Pm >	OL >	Cb
	(-56.30%)	(45.94%)	(-35.80%)	(-35.21%)	(-30.31%)	(-28.18%)
270th day	Spc >	Pm >	CC >	Hc >	OL >	Cb
	(-36.95%)	(-33.42%)	(-22.39%)	(-21.02%)	(-20.57%)	(-19.96%)
300th day	Pm >	CC >	Spc >	Hc >	OL >	Cb
	(-34.53%)	(-32.17%)	(-30.99%)	(-29.30%)	(-27.99%)	(-22.61%)
330th day	Pm >	Spc >	Hc >	CC >	Cb >	OL
	(-30.80%)	(-23.93%)	(-18.15%)	(-18.11%)	(-17.78%)	(-17.28%)
360th day	Pm >	Spc >	Hc >	Cb >	OL >	CC
	(-26.35%)	(-17.05%)	(-14.86%)	(-12.82%)	(-10.69%)	(-10.18%)

DISCUSSION

In the present study, the levels of various constituents of aminergic system Viz. Dopamine (DA), Norepinephrine (NE), Epinephrine (EP), 5-hydroxytryptamine (5-HT) and Monoamine oxidase (MAO) were estimated in different areas of mice brain under prolonged exposure to one of the memory enhancing drug galantamine hydrobromide. Dopamine, Norepinephrine, Epinephrine and Serotonin were elevated in all selected regions upto certain period and after that they exhibited gradual recovery. Contrary to amines, Monoamine oxidase was inhibited in all brain regions for some period and after that the Monoamine oxidase levels showed gradual recovery.

All the drugs, which are used as pharmacological therapies for Alzheimer 's disease are intended to slow the progression of disease. The American Psychiatric Association (2007) recommends the cholinesterase inhibitors such as donepezil, galantamine and rivastigmine for mild to moderate Alzheimer's disease and further suggests that they may be helpful for patients with severe disease also. To date, numerous review articles have been published that summarize the clinical efficacy and safety of drugs for the treatment of Alzheimer's disease (**Geldmacher, 2003, 2007; Lanctot *et al.*, 2003a, 2003b; Trinh *et al.*, 2003; Masterman, 2004; Ritchie *et al.*, 2004; Forchetti, 2005; Birks, 2006; Loveman *et al.*, 2006; Loy and Schneider, 2006; Schmitt *et al.*, 2006; Takeda *et al.*, 2006; Beier, 2007; Hansen *et al.*, 2007**). Most reviews have focused on the second-generation cholinesterase inhibitors (i.e., donepezil, galantamine and rivastigmine), because these drugs mainly target cholinergic and aminergic systems. These review reports lend support to our observations on to the changes in the catecholamines after administration of galantamine into mice for a prolonged period.

There was also evidence that the prolonged usage of antipsychotic drugs which are used to treat schrizophrenia are known to improve dopamine content because of up-regulation of D_2^{High} receptors (**Seeman *et al.*, 2006**). Galantamine is also used sometimes to treat psychosis because the depression was also one of the main reasons for Alzheimer's disease. Results of the present study revealed that there was also elevation in the dopamine content upto around 180 days in all regions of brain with the prolonged usage of galantamine hydrobromide. In this regard, clozapine and quetiopine induce the deviation of D_2^{High} receptors, in contrast to the elevation elicited by haloperidol, risperidone, ziprasidone and olanzapine.

Decreased dopaminergic activity in the cortex of patients with Schrizophrenia and Alzheimer's disease has been inferred on the basis of cerebrospinal fluid and imaging studies (**Weinberger *et al.*, 1988**). However, altered dopamine D_1 receptor density has been related to impaired working memory (**Okubo *et al.*, 1997; Abi-Dargham *et al.*, 2002**). Thus, it has been suggested that cognitive deficits up to 50% in these diseases may be related, in part, to diminished cortical dopaminergic or noradrenergic activity or both (**Meltzer and Mc Gurt, 1999**). To correlate this relationship between the cognitive functions and aminergic system in the present study, elevation of dopamine content in the animals treated with galantamine hydrobromide, one of the memory enhancing drugs was observed. Further support to the present study comes from the evidences that after long and short term administration of memory enhancing drugs like olanzapine and zotepine an increase in dopamine efflux in the cortex occurs (**Kuroki *et al.*, 1999, Rowly *et al.*, 2000; Zhang *et al.*, 2000**). Not only Dopamine, there is also an increase in cortical norepinephrine efflux in prefrontal cortex under the effect of memory enhancing drugs (**Li *et al.*, 1998; Rowley *et al.*, 1998; Zhang *et al.*, 2000**). Thus all memory enhancing drugs are shown to enhance the extracellular 5-HT levels in the median prefrontal cortex and the nucleus accumbans of rats, respectively (**Hertel *et al.*, 1996; Ichikawa *et al.*, 1998**). These earlier reports support all our observations in the present study.

Indeed, several neurological diseases including dementia with Lewy bodies involve dementia with an extra pyramidal disorder and are characterized by a reduction of striatal acetylcholine and dopamine levels as well as a parallel reduction in their striatal receptors (**Piggott *et al.*, 1999, 2003**). For this reason, the currently used pharmacological therapy for dementia with Lewy bodies includes L-dopa and cholinesterase inhibitors, which stabilize cognitive and psychotic symptoms (**Mosimann and Mc Keith, 2003; Kaufer, 2004**). These novel drugs were developed recently inorder to maintain the neuroprospective and monoamine oxidase inhibitory properties (**Youdim and Weinstock, 2002**).

In contrast to spontaneous behaviour, nonselective monoamine oxidase inhibitors in rats and mice have been shown to induce stereotyped hyperactivity in response to L-dopa and L-tryptophan resulting from increased synthesis and release of dopamine and serotonin respectively (**Grahame Smith, 1971; Green and Youdim, 1975; Green *et al.*, 1977**). Several investigators have shown that learning memory can be modified by drugs which affect the central dopamine (DA) neuronal system (**Kesner *et al.*, 1981; Bracs *et al.*, 1984**). So the drugs which are used to improve learning and memory (or) treat Alzheimer's disease could affect the central dopamine neuronal system. It is also well known that psychiatric disorders such as depression and schrizophrenia causing emotional and cognitive disorders are related to an alteration of the serotonergic system (**Dean, 2001; Manji *et al.*, 2001; Pralong *et al.*, 2002; Trudeau, 2004**).

Alzheimer's disease, with severe cognitive disorders and cholinergic deficits is characterized by altered levels of glutamate and serotonin as well as their receptors in brain regions involved in learning and memory (**D'Amanto *et al.*, 1987; Jansen *et al.*, 1990; Nordberg, 1992; Doraiswamy, 2003**). Again it has been shown that glutamate transmission, which is defective in Alzheimer's disease, can be pharmacologically modulated using antagonists of 5-HT_{1A} receptors in Cerebral Cortex (**Bowen *et al.*, 1992; Francis *et al.*, 1993**) and in the hippocampus suggesting that also Alzheimer's disease patients may be benefited from a treatment with serotonergic drugs.

In the present study also, depending upon the 5-HT_{1A} receptors, the serotonin levels were elevated differentially in different regions of mice brain when treated with galantamine for prolonged exposure. The drug, galantamine hydrobromide selected in the present study is a tertiary alkaloid with a proposed dual mode of action: competitive inhibition of acetylcholinesterase and allosteric modulation of nicotine receptors (**Bores *et al.*, 1996**). There was also evidence that the nicotine receptor modulators act as the aminergic system neurotransmitter enhancers. This statement was proved true in the present study where elevated levels of monoamines were registered in mice brain under the effect of galantamine hydrobromide.

Direct studies have revealed that it is likely that loss and atrophy of the neurons are responsible for the deficiencies in brain noradrenaline content (**Adolfsson *et al.*,**

1979; Mann *et al.*, 1980 and Yates *et al.*, 1981) dopamine – β – hydroxylase activity **(Cross *et al.*, 1981)** and MHPG levels in brain **(Perry *et al.*, 1981)** and urine **(Mann *et al.*, 1980)** leading to Alzheimer's disease. Furthermore, a greater loss of nerve cells from the locus caeruleus seems to occur in those cases of Alzheimer's disease with high plaque counts **(Tomilson *et al.*, 1981)** and is related to a greater degree of mental impairment **(Bondarff *et al.*, 1981)**. These findings indicate that early changes in the noradrenaline neurotransmitter system may play a fundamental role in the pathogenesis of Alzheimer's disease. There are reports on the clinical trials of levodopa where the patients with Alzheimer's disease but without extrapyramidal signs showed significant improvements in mental performance during treatment for over periods of 6 to 9 months which were lost during a drug free period, but subsequently regained on resumption of medication **(Lewis *et al.*, 1978; Johnson *et al.*, 1978; Jellinger *et al.*, 1980)**. Since the dopamine system in brain is essentially unaltered in Alzheimer's disease **(Davies *et al.*, 1976; Davies, 1979; Adolfsson *et al.*, 1979)**, it is therefore possible that beneficial effects of L-dopa on mental function may arise through preferential modulation (stimulation) of the noradrenaline system. So, the drugs which are used to treat Alzheimer's disease also show their direct positive effect on noradrenaline system and improve the noradrenaline and dopamine levels **(Mann *et al.*, 1980)**. This report further strengthen our observations in the present study where administration of galantamine caused elevation in 5-HT and noradrenaline levels in mice in the absence of disease.

Numerous therapeutic targets have been pursued in order to develop agents for cognitive enhancement in the CNS diseases like Alzheimer's disease and Schrizophrenia **(Robichaud, 2006)**. As one of these therapeutic targets, the 5-HT_6 receptor which is one of the most recently discovered 5-HT receptor subtypes was first cloned in 1993 **(Monsma *et al.*, 1993; Ruat *et al.*, 1993)**. Numerous invivo studies have shown that blockade of 5-HT_6 receptor function improves cognition in a number of rodent models. In addition, invivo microdialysis studies have shown that 5-HT_6 receptor antagonism enhances neurotransmission at cholinergic and glutamatergic neurons, as well as in other pathways. Therefore, it can be inferred that antagonism of the 5-HT_6 receptor can potentially provide an effective treatment for cognitive impairment in Alzheimer's disease and Schrizophrenia and has been the subject of intense research **(Woolley *et al.*,**

2009; Glennon, 2003; Holenz *et al.*, 2006; Liu and Robichaud, 2009). So, nowadays the pharmacological compounds are synthesized on the basis of that the compounds block the 5-HT_6 receptors and improves neurotransmitter amounts to treat central nervous system diseases like Alzheimer's disease **(Keivin *et al.*, 2009).**

It was also clear that the Alzheimer's patients suffer with low levels of choline and amine neurotranmitters. For these reasons, currently acetylcholinesterase inhibitors are the drugs approved for treatment of cognitive dysfunction in Alzheimer's disease **(Millard *et al.*, 1995 and Giacobini, 2001).** But acetylcholine enhancing drugs can compensate for only part of the neuronal dysfunction in Alzheimer's disease. Among the drugs now moving into clinical trials are several compounds that may modify the progression of Alzheimer's disease. Apart from these, another potential therapy for Alzheimer's disease is the use of MAO-B inhibitors **(Reiderer *et al.*, 2004).** All cholinesterase inhibitors are also act as selective MAO-inhibitors **(Gokhan *et al.*, 2003)** of human erythrocytes and plasma. So, it is proved that these are used for treatment of cognitive dysfunction in all areas selectively and none competitively **(Ucar *et al.*, 2005).** This particular report strongly support the present investigation where the selected acetylcholine inhibitor, galantamine also inhibited MAO activity in all regions of mice brain.

Degeneration of the locus coeruleus **(Berger, 1984)**, the source of noradrenergic innervations to the cerebral cortex and hippocampus **(Marciniuk *et al.*, 1986)** compromises noradrenergic functioning in people with Alzheimer's disease. Pre-clinical evidence links the noradrenergic system to specific spheres in learning and memory **(Randt *et al.*, 1971; Kety, 1972; Stein *et al.*, 1975; Haroutunian *et al.*, 1986; Mohr *et al.*, 1989b).** Attempts to improve the cognition of patients with Alzheimer's disease using alpha-2 receptor agonists have been disappointing **(Mohr *et al.*, 1989b; Schlegel *et al.*, 1989).** Since, currently available compounds again lack a high degree of receptor specificity and in all likelihood act simultaneously on other symptoms in ways that are frequently not understood, the concept of noradrenergic enhancement needs to be explored further with more precise pharmacological tools. For this reason, the recently inventing drugs are made up with these functions showing activities and their chemical structures are also having these noradrenergic enhancing capacities.

Disturbances in serotonergic mechanisms have been linked to several of the behavioral consequences of Alzheimer's disease (**Quirion *et al.*, 1986**). Reductions in serotonin (5-HT) markers have consistently been demonstrated, particularly in cortical regions. So, the pharmacological drugs which are used to treat Alzheimer's disease having the capacity to involve re-uptake inhibition with 5-HT in patients have suggested improvement in behavioral disturbances (**Nyth and Gottfries, 1990**). These drugs show reductions in energy and depression related symptoms in patients with Alzheimer's disease (**Lawlor *et al.*, 1991**). Because of these reasons our selected memory enhancing drug galantamine which is used to treat Alzheimer's disease shows elevated levels of 5-HT in absence of alzheimer's conditions also.

The above mentioned earlier reports say without doubt that conceptual improvements in drug development will enhance the chances for success. Future traces of antidementia compounds might include matching individual patient's transmitter deficits with a particular drug's pharmacological action ("treating toward the deficit") (**Mohr and Chase, 1991**). The enhancement of cholinergic function remains the most successful approach to date for ameliorating the symptoms of Alzheimer's disease. This strategy is based on the cholinergic hypothesis (**Francis *et al.*, 1999**) which proposes that degeneration of cholinergic neurons in the basal fore brain and the associated loss of cholinergic neurotransmission in the Cerebral Cortex contribute significantly to the cognitive decline seen in patients with Alzheimer's disease. Acetylcholinesterase inhibition is currently the most established strategy for correcting cholinergic deficits in Alzheimer's disease and improves cognitive symptoms (**Schneider, 1996; Nordberg and Evensson, 1998**). Galantamine is a novel agent that modulates nicotinic receptors (**Schrattenholz *et al.*, 1996 and Albuquerque *et al.*, 1997**) and potentiates nicotinic neurotransmission in addition to inhibiting acetylcholinesterase (**Bores *et al.*, 1996**). Given the loss of nicotinic receptors that accompanies the impairment of presynaptic cholinergic function in Alzheimer's disease (**Nordberg *et al.*, 1989**) and their role in memory and learning (**Newhouse *et al.*, 1997**) maintaining nicotinic activity may have therapeutic value (**Albuquerque *et al.*, 1997**).

This enhancement of nicotinic neurotransmitter may be clinically relevant because activation of presynaptic nicotinic receptors has been shown to increase the release of acetylcholine, catecholamines and glutamate which are deficit in Alzheimer's disease and are thought to be involved in memory and learning (**Francis *et al.*, 1999**). The present selected drug galantamine can also have the property that crosses the blood-brain-barrier (**Corey-Bloom, 2003**).

Not surprisingly, MAO-A which is often targeted for the treatment of depression, is also a potential risk factor for late onset of Alzheimer's disease (**Emilson *et al.*, 2002; Taketashi *et al.*, 2002 and Nishimura *et al.*, 2005**). Inhibition of MAO-A activity protects against striatal damage produced by the mitochondrial poison malonate and appears to rely on attenuation of dopamine derived ROS (**Maragos *et al.*, 2004**) while apoptosis following serum starvation is reduced in MAO-A deficient cortical brain cells (**Ou *et al.*, 2006**). So, the apoptosis or neurodegeneration of nerve cells during Alzheimer's disease was protected by the drugs showing MAO inhibition property. To the correlation for this in the present study the selected drug galantamine shows the MAO inhibition property in mice. And also, there is evidence that the normal aging process such as Parkinson's disease (**Jenner and Olanow, 1996**) and Alzheimer's disease (**Kennedy *et al.*, 2003**) could well be affected by oxidative stress associated with enhanced MAO-generated levels of H_2O_2. So, the drugs treated for Alzheimer's disease shows, decreased levels of the MAO activity.

There were also studies on the basis of relation between serotonin and dopamine that is serotonin can alter dopaminergic signal transmission in several ways. For example, by interacting with the 5-HT_2 receptor, serotonin stimulates the activity of dopaminergic neurons in the brain region called the Ventral Tegmental Area (VTA), thereby causing increased dopamine release (**Campbell *et al.*, 1996**). Serotonin release in the brain regions can stimulate dopamine release, presumably by activationg 5-HT3 receptors located on the endings of dopaminergic neurons (**Grant, 1995; Campbell and Bride,**

1991). This is the best support to the present study, here also the elevated levels of serotonin shows the elevation in dopamine content also in different regions of mice brain.

In general, the present investigation on biogenic amines, their metabolites and MAO activity in different areas of mice brain following administration of galantamine, has shown that galantamine steped up the accumulation of biogenic amines in all the brain areas, while it caused a decrease in MAO activity in all areas. These results were in general agreement with those obtained with other researches and other memory enhancing or Alzheimer's disease treated pharmacological drugs with some deviations. The effect of oral administration (5mg/Kg) body weight of galantamine used in the present study, the maximal positive peaks for Dopamine, Norepinephrine, Epinephrine and 5-Hydroxytryptophan obtained at 165 to 210 days and return of the parameters to control levels by about 360 days after the administration of galantamine appear reasonable. From correlated to the above the MAO activity was also inhibited upto 195 days and after that it was also reached to control levels. From this observation it was concluded that the effect of galantamine on diseased or non diseased persons in the biogenic amines are the same. The usage of memory enhancing drugs like galantamine for the improvement of memory or intelligence by the normal persons shows some what positive effect upto certain period after that it shows no effects to improve their memory and the memory attains normal.

On Aminergic system, it is concluded that galantamine caused very significant disturbances in the entire metabolic pathway of the aminergic system on different days and to different extent. The reason for different regions of the brain exhibiting different levels of sensitivity to galantamine hydrobromide treatment perhaps may be because of the heterogeneity of the brain and also the functions with which they are associated. For example olfactory lobe is implicated in the manifestation of sensory functions, hippocampus in memory and cognitive function, cerebral cortex in reflex actions and locomotor activities, cerebellum in locomotory functions and co-ordination of movements, pons medulla in respiratory functions and spinal cord in voluntary movements and co-ordination of muscles.

Table 4.1 : Changes in Dopamine Content (µg/g wet wt) in Olfactory Lobe (OL), Cerebellum (CB), Cerebral Cortex (CC) Hippocampus(Hc), Pons medulla (Pm) and Spinal Cord (SPC) of male albino mice on prolonged exposure to Galantamine Hydrobromide. Values in parantheses indicate percent changes from control.

CHRONIC																								
	C	15 DAYS	C	30 DAYS	C	45 DAYS	C	60 DAYS	C	75 DAYS	C	90 DAYS	C	105 DAYS	C	120 DAYS	C	135 DAYS	C	150 DAYS	C	165 DASYS	C	180 DAYS
OL	0.256 ±0.01	0.325** ±0.01 (26.95)	0.221 ±0.004	0.289** ±0.004 (30.76)	0.203 ±0.004	0.268** ±0.01 (32.01)	0.246 ±0.02	0.362** ±0.01 (47.15)	0.216 ±0.01	0.321** ±0.008 (48.61)	0.265 ±0.01	0.398** ±0.02 (50.18)	0.231 ±0.01	0.352** ±0.008 (52.38)	0.222 ±0.01	0.356** ±0.008 (60.36)	0.216 ±0.004	0.354** ±0.01 (63.88)	0.241 ±0.01	0.401** ±0.02 (66.39)	0.213 ±0.04	0.369** ±0.02 (73.23)	0.211 ±0.01	0.396** ±0.02 (87.67)
CB	0.416 ±0.01	0.524** ±0.01 (25.96)	0.409 ±0.006	0.558** ±0.01 (36.43)	0.406 ±0.01	0.567** ±0.01 (39.65)	0.402 ±0.01	0.598** ±0.01 (48.75)	0.410 ±0.01	0.612** ±0.02 (49.26)	0.425 ±0.01	0.642** ±0.008 (51.05)	0.436 ±0.01	0.684** ±0.01 (56.88)	0.418 ±0.008	0.662** ±0.008 (58.37)	0.431 ±0.008	0.698** ±0.01 (61.94)	0.413 ±0.01	0.687** ±0.01 (66.34)	0.432 ±0.01	0.712** ±0.01 (64.81)	0.405 ±0.004	0.685** ±0.01 (69.13)
HC	0.615 ±0.02	0.816** ±0.008 (32.68)	0.631 ±0.01	0.882** ±0.02 (39.77)	0.639 ±0.02	0.895** ±0.01 (40.06)	0.613 ±0.01	0.865** ±0.01 (41.10)	0.627 ±0.01	0.896** ±0.01 (42.90)	0.632 ±0.02	0.916** ±0.004 (44.93)	0.617 ±0.01	0.902** ±0.01 (46.19)	0.635 ±0.01	0.958** ±0.008 (50.86)	0.641 ±0.01	0.968** ±0.04 (51.01)	0.623 ±0.01	0.968** ±0.008 (55.37)	0.621 ±0.01	0.995** ±0.08 (60.22)	0.662 ±0.01	1.101** ±0.03 (66.31)
CC	0.785 ±0.008	0.921** ±0.004 (17.32)	0.715 ±0.004	0.900** ±0.01 (25.99)	0.726 ±0.008	0.931** ±0.008 (28.23)	0.732 ±0.008	0.962** ±0.008 (31.42)	0.708 ±0.01	0.954** ±0.008 (34.74)	0.718 ±0.01	0.996** ±0.02 (38.71)	0.705 ±0.004	0.974** ±0.008 (38.15)	0.713 ±0.008	1.010** ±0.04 (41.65)	0.733 ±0.008	0.998** ±0.04 (42.36)	0.718 ±0.008	1.102** ±0.04 (55.43)	0.698 ±0.008	1.115** ±0.04 (59.74)	0.709 ±0.01	1.210** ±0.08 (65.07)
PM	0.501 ±0.03	0.625** ±0.01 (24.75)	0.505 ±0.01	0.654** ±0.008 (29.50)	0.512 ±0.008	0.698** ±0.01 (36.32)	0.518 ±0.01	0.712** ±0.008 (37.45)	0.518 ±0.02	0.718** ±0.01 (38.61)	0.506 ±0.01	0.715** ±0.004 (41.30)	0.546 ±0.01	0.794** ±0.008 (45.42)	0.521 ±0.01	0.756** ±0.01 (45.10)	0.525 ±0.01	0.776** ±0.01 (47.80)	0.531 ±0.01	0.789** ±0.008 (48.58)	0.513 ±0.01	0.785** ±0.01 (53.02)	0.526 ±0.008	0.796** ±0.008 (51.33)
SPC	0.315 ±0.01	0.365** ±0.004 (15.87)	0.306 ±0.008	0.398** ±0.02 (30.06)	0.321 ±0.008	0.426** ±0.01 (32.71)	0.317 ±0.01	0.425** ±0.008 (34.06)	0.301 ±0.01	0.415** ±0.01 (37.87)	0.313 ±0.01	0.456** ±0.01 (45.68)	0.336 ±0.008	0.508** ±0.008 (51.19)	0.339 ±0.01	0.528** ±0.01 (55.75)	0.326 ±0.008	0.516** ±0.01 (58.28)	0.352 ±0.008	0.569** ±0.008 (61.64)	0.319 ±0.008	0.525** ±0.01 (64.57)	0.341 ±0.01	0.560** ±0.01 (64.22)

Values are mean ± SEM of six observations each from tissues pooled from 6 animals

**Values are significant at $p<0.01$

* Indicate significance at $p<0.05$

Not significant

Table 4.2 : Changes in Dopamine Content (μg/g wet wt) in Olfactory Lobe (OL), Cerebellum (CB), Cerebral Cortex (CC) Hippocampus(Hc), Pons medulla (Pm) and Spinal Cord (SPC) of male albino mice on prolonged exposure to Galantamine Hydrobromide. Values in parantheses indicate percent changes from control.

	CHRONIC																							
	C	195 DAYS	C	210 DAYS	C	225 DAYS	C	240 DAYS	C	255 DAYS	C	270 DAYS	C	285 DAYS	C	300 DAYS	C	315 DAYS	C	330 DAYS	C	345 DASYS	C	360 DAYS
OL	0.211 ± 0.004	0.368** ± 0.01 (74.40)	0.224 ±0.01	0.365** ±0.01 (62.94)	0.221 ±0.01	0.358** ±0.01 (61.99)	0.235 ±0.01	0.369** ±0.01 (57.02)	0.208 ±0.01	0.325** ±0.008 (56.25)	0.206 ±0.004	0.315** ±0.008 (52.91)	0.216 ±0.004	0.325** ±0.008 (50.46)	0.218 ±0.008	0.325** ±0.01 (49.08)	0.254 ±0.008	0.357** ±0.008 (40.55)	0.231 ±0.01	0.315** ±0.01 (36.36)	0.242 ±0.008	0.301** ±0.01 (24.38)	0.226 ±0.01	0.325** ±0.01 (22.56)
CB	0.418 ±0.004	0.706** ±0.008 (68.89)	0.444 ±0.01	0.718** ±0.004 (61.7)	0.456 ±0.01	0.722** ±0.01 (58.33)	0.416 ±0.01	0.642** ±0.01 (54.32)	0.432 ±0.01	0.665** ±0.01 (53.93)	0.414 ±0.01	0.632** ±0.004 (52.65)	0.422 ±0.01	0.625** ±0.008 (48.10)	0.436 ±0.01	0.598** ±0.008 (37.15)	0.434 ±0.008	0.599** ±0.01 (37.38)	0.413 ±0.01	0.568** ±0.01 (37.53)	0.419 ±0.01	0.552** ±0.01 (31.74)	0.418 ±0.01	0.532** ±0.008 (27.27)
HC	0.632 ±0.01	0.996** ±0.01 (57.59)	0.618 ±0.008	0.969** ±0.01 (56.79)	0.628 ±0.01	0.968** ±0.02 (54.14)	0.636 ±0.01	0.958** ±0.01 (50.62)	0.614 ±0.01	0.899** ±0.02 (46.41)	0.637 ±0.01	0.912** ±0.01 (43.17)	0.624 ±0.01	0.890** ±0.01 (42.62)	0.651 ±0.008	0.869** ±0.01 (33.48)	0.615 ±0.008	0.810** ±0.01 (31.70)	0.653 ±0.01	0.852** ±0.01 (30.47)	0.629 ±0.01	0.735** ±0.004 (16.85)	0.621 ±0.01	0.698** ±0.01 (12.39)
CC	0.721 ±0.01	1.180** ±0.04 (61.20)	0.701 ±0.008	1.121** ±0.03 (59.91)	0.726 ±0.01	1.117** ±0.04 (53.85)	0.715 ±0.008	1.012** ±0.04 (41.53)	0.711 ±0.008	0.998** ±0.02 (40.36)	0.713 ±0.01	0.998** ±0.02 (39.97)	0.728 ±0.01	0.991** ±0.02 (36.13)	0.721 ±0.01	0.965** ±0.008 (33.84)	0.718 ±0.008	0.943** ±0.01 (31.33)	0.715 ±0.01	0.913** ±0.01 (27.19)	0.697 ±0.008	0.869** ±0.01 (24.67)	0.703 ±0.01	0.869** ±0.01 (23.61)
PM	0.516 ±0.004	0.779** ±0.02 (50.96)	0.531 ±0.01	0.796** ±0.01 (49.90)	0.541 ±0.01	0.789** ±0.01 (45.84)	0.518 ±0.008	0.748** ±0.004 (44.40)	0.514 ±0.01	0.742** ±0.01 (44.35)	0.532 ±0.01	0.751** ±0.01 (41.16)	0.512 ±0.01	0.715** ±0.01 (39.64)	0.512 ±0.01	0.705** ±0.01 (37.69)	0.515 ±0.01	0.701** ±0.02 (36.11)	0.518 ±0.01	0.645** ±0.008 (24.51)	0.498 ±0.01	0.615** ±0.01 (23.69)	0.526 ±0.008	0.635** ±0.008 (20.72)
SPC	0.332 ±0.008	0.589** ±0.008 (77.40)	0.316 ±0.01	0.512** ±0.01 (62.02)	0.312 ±0.01	0.490** ±0.02 (56.65)	0.334 ±0.01	0.516** ±0.008 (52.66)	0.301 ±0.004	0.447** ±0.01 (48.50)	0.318 ±0.01	0.458** ±0.01 (44.02)	0.326 ±0.008	0.469** ±0.02 (43.86)	0.313 ±0.01	0.414** ±0.01 (32.26)	0.316 ±0.01	0.412** ±0.004 (30.37)	0.311 ±0.01	0.398** ±0.01 (27.97)	0.312 ±0.008	0.369** ±0.008 (18.26)	0.308 ±0.01	0.356* ±0.01 (15.58)

Values are mean ± SEM of six observations each from tissues pooled from 6 animals

**Values are significant at $p<0.01$

* Indicate significance at $p<0.05$

Not significant

Table 5.1 : Changes in Norepinephrine Content (µg/g wet wt) in Olfactory Lobe (OL), Cerebellum (CB), Cerebral Cortex (CC) Hippocampus(Hc), Pons medulla (Pm) and Spinal Cord (SPC) of male albino mice on prolonged exposure to Galantamine Hydrobromide. Values in parantheses indicate percent changes from control.

	CHRONIC																							
	C	15 DAYS	C	30 DAYS	C	45 DAYS	C	60 DAYS	C	75 DAYS	C	90 DAYS	C	105 DAYS	C	120 DAYS	C	135 DAYS	C	150 DAYS	C	165 DASYS	C	180 DAYS
OL	0.441 ± 0.02	0.562** ± 0.01 (27.43)	0.427 ±0.02	0.568** ±0.01 (33.02)	0.438 ±0.01	0.612** ±0.01 (39.72)	0.416 ±0.01	0.598** ±0.02 (43.75)	0.418 ±0.01	0.612** ±0.008 (46.41)	0.426 ±0.01	0.635** ±0.01 (49.06)	0.401 ±0.01	0.602** ±0.01 (50.12)	0.416 ±0.01	0.639** ±0.02 (53.60)	0.430 ±0.01	0.674** ±0.02 (56.74)	0.432 ±0.01	0.685** ±0.02 (58.56)	0.411 ±0.01	0.658** ±0.02 (60.09)	0.427 ±0.01	0.698** ±0.02 (63.46)
CB	0.822 ±0.02	0.985** ±0.02 (19.82)	0.813 ±0.02	0.978** ±0.01 (20.29)	0.816 ±0.02	1.021** ±0.03 (25.12)	0.831 ±0.01	1.102** ±0.02 (32.61)	0.836 ±0.04	1.125** ±0.04 (34.56)	0.826 ±0.04	1.127** ±0.02 (36.44)	0.869 ±0.05	1.212** ±0.08 (39.47)	0.842 ±0.03	1.225** ±0.05 (45.48)	0.815 ±0.02	1.216** ±0.02 (49.20)	0.820 ±0.02	1.240** ±0.03 (51.21)	0.833 ±0.02	1.331** ±0.03 (59.78)	0.816 ±0.02	1.321** ±0.03 (61.88)
HC	1.265 ±0.02	1.621** ±0.05 (21.96)	1.268 ±0.04	1.612** ±0.05 (27.12)	1.524 ±0.03	2.012** ±0.08 (32.02)	1.215 ±0.04	1.625** ±0.05 (33.74)	1.118 ±0.02	1.552** ±0.03 (38.08)	1.123 ±0.008	1.568** ±0.05 (39.62)	1.125 ±0.02	1.658** ±0.04 (47.37)	0.981 ±0.03	1.485** ±0.05 (51.37)	1.315 ±0.02	2.021** ±0.17 (53.68)	1.105 ±0.01	1.858** ±0.06 (68.14)	1.218 ±0.03	2.081** ±0.07 (70.85)	1.318 ±0.04	2.358**# ±0.13 (78.90)
CC	1.712 ±0.07	2.121** ±0.10 (23.89)	1.814 ±0.03	2.457** ±0.28 (35.64)	1.514 ±0.13	2.112** ±0.19 (39.65)	1.212 ±0.10	1.756** ±0.04 (44.88)	1.515 ±0.04	2.251** ±0.08 (48.58)	1.218 ±0.09	1.854** ±0.03 (52.30)	1.216 ±0.09	1.898** ±0.02 (56.08)	1.316 ±0.11	2.112** ±0.10 (60.48)	1.118 ±0.02	1.852** ±0.04 (65.65)	1.451 ±0.08	2.458** ±0.12 (69.40)	1.653 ±0.03	2.861** ±0.17 (73.07)	1.531 ±0.12	2.658** ±0.18 (73.61)
PM	2.435 ±0.20	3.112* ±0.27 (28.08)	2.313 ±0.25	3.101* ±0.17 (34.06)	2.181 ±0.09	3.011** ±0.15 (38.05)	2.225 ±0.28	3.213* ±0.17 (44.40)	2.516 ±0.17	3.698** ±0.20 (46.90)	2.185 ±0.09	3.215** ±0.17 (47.10)	2.179 ±0.28	3.339** ±0.11 (53.24)	2.193 ±0.09	3.458** ±0.22 (57.68)	2.178 ±0.28	3.581** ±0.17 (64.40)	2.121 ±0.13	3.552** ±0.18 (67.40)	2.318 ±0.25	4.012** ±0.25 (73.08)	2.286 ±0.21	3.981** ±0.24 (74.18)
SPC	0.639 ±0.01	0.751** ±0.01 (17.52)	0.641 ±0.02	0.758** ±0.02 (18.25)	0.650 ±0.01	0.798** ±0.01 (22.76)	0.601 ±0.03	0.756** ±0.02 (25.79)	0.632 ±0.02	0.812** ±0.01 (28.48)	0.617 ±0.01	0.825** ±0.03 (33.71)	0.612 ±0.03	0.824** ±0.02 (34.64)	0.640 ±0.01	0.878** ±0.02 (37.18)	0.611 ±0.02	0.856** ±0.02 (40.09)	0.626 ±0.02	0.899** ±0.02 (43.61)	0.600 ±0.02	0.902** ±0.01 (50.33)	0.603 ±0.01	0.897** ±0.01 (48.75)

Values are mean ± SEM of six observations each from tissues pooled from 6 animals

**Values are significant at $p<0.01$

* Indicate significance at $p<0.05$

Not significant

Table 5.2 : Changes in Norepinephrine Content (μg/g wet wt) in Olfactory Lobe (OL), Cerebellum (CB), Cerebral Cortex (CC) Hippocampus(Hc), Pons medulla (Pm) and Spinal Cord (SPC) of male albino mice on prolonged exposure to Galantamine Hydrobromide. Values in parantheses indicate percent changes from control.

CHRONIC																								
	C	195 DAYS	C	210 DAYS	C	225 DAYS	C	240 DAYS	C	255 DAYS	C	270 DAYS	C	285 DAYS	C	300 DAYS	C	315 DAYS	C	330 DAYS	C	345 DASYS	C	360 DAYS
OL	0.427 ±0.01	0.687** ±0.02 (60.88)	0.429 ±0.02	0.677** ±0.04 (57.80)	0.421 ±0.02	0.654** ±0.02 (55.34)	0.415 ±0.01	0.637** ±0.02 (53.49)	0.433 ±0.01	0.645** ±0.02 (48.96)	0.437 ±0.01	0.635** ±0.01 (45.30)	0.414 ±0.02	0.598** ±0.01 (44.64)	0.456 ±0.01	0.652** ±0.01 (42.98)	0.401 ±0.02	0.565** ±0.01 (40.89)	0.428 ±0.02	0.569** ±0.01 (32.94)	0.461 ±0.01	0.599** ±0.02 (29.93)	0.452 ±0.02	0.579** ±0.01 (28.09)
CB	0.841 ±0.02	1.325** ±0.02 (57.55)	0.843 ±0.01	1.321** ±0.03 (56.70)	0.826 ±0.01	1.255** ±0.02 (51.93)	0.817 ±0.02	1.201** ±0.02 (47.00)	0.835 ±0.02	1.215** ±0.02 (45.50)	0.833 ±0.01	1.210** ±0.02 (45.25)	0.855 ±0.01	1.212** ±0.03 (41.75)	0.813 ±0.02	1.219** ±0.004 (38.86)	0.811 ±0.02	1.111** ±0.02 (36.99)	0.826 ±0.004	1.021** ±0.01 (23.60)	0.816 ±0.004	0.998** ±0.003 (22.30)	0.828 ±0.02	0.999** ±0.03 (20.65)
HC	1.212 ±0.02	2.132** ±0.06 (75.90)	1.189 ±0.04	1.998** ±0.10 (68.00)	1.215 ±0.02	2.010** ±0.08 (65.43)	1.015 ±0.04	1.658** ±0.01 (63.34)	1.712 ±0.02	2.695** ±0.09 (57.41)	1.516 ±0.01	2.323** ±0.19 (53.23)	1.418 ±0.01	2.099** ±0.07 (47.9)	1.400 ±0.01	2.012** ±0.13 (43.71)	1.216 ±0.02	1.658** ±0.01 (36.28)	1.312 ±0.03	1.758** ±0.04 (33.99)	1.261 ±0.02	1.615** ±0.04 (28.07)	1.161 ±0.02	1.468** ±0.04 (26.44)
CC	1.138 ±0.02	2.010** ±0.08 (76.6)	1.261 ±0.11	2.126** ±0.10 (68.5)	1.678 ±0.02	2.678** ±0.19 (59.59)	1.125 ±0.09	1.785** ±0.02 (58.66)	1.314 ±0.12	2.010** ±0.08 (49.75)	1.129 ±0.09	1.654** ±0.04 (47.20)	1.501 ±0.10	2.125** ±0.10 (41.57)	1.426 ±0.10	1.989** ±0.10 (39.48)	1.543 ±0.04	2.120** ±0.10 (37.39)	1.241 ±0.09	1.658** ±0.04 (33.60)	1.229 ±0.08	1.541* ±0.08 (25.38)	1.331 ±0.11	1.659* ±0.04 (24.64)
PM	2.282 ±0.10	3.783** ±0.21 (65.7)	2.365 ±0.27	3.612** ±0.22 (61.2)	2.116 ±0.17	3.331** ±0.11 (57.41)	2.212 ±0.16	3.451** ±0.17 (56.01)	2.176 ±0.23	3.252** ±0.15 (49.4)	2.182 ±0.09	3.151** ±0.16 (44.4)	2.238 ±0.15	3.122** ±0.17 (40.7)	2.315 ±0.25	3.112* ±0.17 (34.4)	2.421 ±0.18	3.218** ±0.16 (32.9)	2.413 ±0.18	3.118* ±0.17 (29.12)	2.518 ±0.18	3.236* ±0.16 (28.51)	2.281 ±0.10	3.012** ±0.15 (26.69)
SPC	0.627 ±0.02	0.899** ±0.02 (43.59)	0.605 ±0.02	0.925** ±0.02 (44.90)	0.617 ±0.01	0.877** ±0.01 (42.13)	0.651 ±0.01	0.911** ±0.02 (39.99)	0.621 ±0.01	0.859** ±0.01 (38.32)	0.612 ±0.01	0.821** ±0.02 (34.15)	0.614 ±0.02	0.810** ±0.008 (31.92)	0.635 ±0.01	0.825** ±0.02 (29.92)	0.601 ±0.02	0.768** ±0.01 (27.78)	0.638 ±0.02	0.785** ±0.02 (22.88)	0.613 ±0.02	0.741** ±0.02 (20.88)	0.639 ±0.02	0.765** ±0.02 (19.71)

Values are mean ± SEM of six observations each from tissues pooled from 6 animals

**Values are significant at $p<0.01$

* Indicate significance at $p<0.05$

Not significant

Table 6.1 : Changes in Epinephrine Content (µg/g wet wt) in Olfactory Lobe (OL), Cerebellum (CB), Cerebral Cortex (CC) Hippocampus(Hc), Pons medulla (Pm) and Spinal Cord (SPC) of male albino mice on prolonged exposure to Galantamine Hydrobromide. Values in parantheses indicate percent changes from control.

CHRONIC																								
	C	15 DAYS	C	30 DAYS	C	45 DAYS	C	60 DAYS	C	75 DAYS	C	90 DAYS	C	105 DAYS	C	120 DAYS	C	135 DAYS	C	150 DAYS	C	165 DASYS	C	180 DAYS
OL	0.252 ± 0.006	0.302** ± 0.004 (19.8)	0.256 ±0.01	0.321** ±0.004 (25.3)	0.235 ±0.01	0.301** ±0.004 (28.0)	0.245 ±0.02	0.321** ±0.004 (31.02)	0.218 ±0.01	0.299** ±0.01 (37.1)	0.255 ±0.01	0.358** ±0.008 (40.3)	0.211 ±0.01	0.301** ±0.01 (42.6)	0.217 ±0.02	0.315** ±0.004 (45.16)	0.242 ±0.008	0.358** ±0.02 (47.9)	0.218 ±0.008	0.325** ±0.01 (49.0)	0.234 ±0.003	0.358** ±0.008 (52.9)	0.220 ±0.01	0.344** ±0.008 (56.3)
CB	0.425 ±0.02	0.625** ±0.02 (16.8)	0.462 ±0.03	0.612** ±0.01 (32.4)	0.455 ±0.01	0.625** ±0.02 (37.3)	0.405 ±0.02	0.569** ±0.02 (40.4)	0.410 ±0.02	0.589** ±0.02 (43.6)	0.432 ±0.01	0.632** ±0.01 (46.2)	0.471 ±0.02	0.698** ±0.01 (48.1)	0.401 ±0.02	0.601** ±0.01 (49.8)	0.438 ±0.02	0.658** ±0.02 (50.2)	0.445 ±0.01	0.678** ±0.02 (52.3)	0.412 ±0.02	0.648** ±0.02 (57.2)	0.433 ±0.01	0.679** ±0.02 (56.8)
HC	0.620 ±0.03	0.765** ±0.02 (22.5)	0.629 ±0.02	0.784** ±0.008 (24.6)	0.620 ±0.02	0.783** ±0.004 (26.2)	0.643 ±0.03	0.834** ±0.04 (29.8)	0.619 ±0.01	0.822** ±0.03 (32.7)	0.613 ±0.004	0.829** ±0.03 (35.2)	0.593 ±0.01	0.814** ±0.02 (37.2)	0.621 ±0.02	0.883** ±0.02 (42.1)	0.605 ±0.02	0.880** ±0.01 (45.4)	0.596 ±0.02	0.876** ±0.01 (46.9)	0.591 ±0.02	0.880** ±0.01 (48.9)	0.589 ±0.02	0.887** ±0.01 (50.5)
CC	0.581 ±0.02	0.720** ±0.01 (23.9)	0.578 ±0.02	0.725** ±0.01 (25.4)	0.578 ±0.02	0.745** ±0.02 (28.8)	0.535 ±0.01	0.712** ±0.01 (33.0)	0.551 ±0.01	0.745** ±0.01 (35.2)	0.568 ±0.02	0.789** ±0.02 (38.9)	0.552 ±0.01	0.784** ±0.02 (42)	0.545 ±0.01	0.798** ±0.01 (46.4)	0.525 ±0.01	0.785** ±0.01 (49.5)	0.526 ±0.008	0.798** ±0.02 (51.7)	0.515 ±0.02	0.798** ±0.02 (54.9)	0.510 ±0.02	0.801** ±0.008 (57.0)
PM	0.695 ±0.004	0.825** ±0.004 (18.7)	0.686 ±0.004	0.842** ±0.003 (22.7)	0.720 ±0.01	0.902** ±0.001 (25.2)	0.722 ±0.008	0.922** ±0.01 (27.7)	0.682 ±0.008	0.875** ±0.008 (28.2)	0.751 ±0.068	0.985** ±0.004 (31.1)	0.710 ±0.01	0.955** ±0.01 (34.5)	0.646 ±0.01	0.886** ±0.004 (37.1)	0.651 ±0.01	0.925** ±0.01 (42.0)	0.634 ±0.01	0.930** ±0.01 (46.2)	0.652 ±0.02	0.975** ±0.004 (49.5)	0.635 ±0.02	0.962** ±0.02 (51.4)
SPC	0.365 ±0.02	0.452* ±0.02 (23.8)	0.318 ±0.02	0.414** ±0.02 (30.1)	0.313 ±0.03	0.415* ±0.01 (32.1)	0.323 ±0.01	0.438** ±0.02 (34.6)	0.346 ±0.02	0.485** ±0.02 (40.1)	0.315 ±0.02	0.451** ±0.02 (43.1)	0.352 ±0.01	0.518** ±0.02 (46.2)	0.325 ±0.02	0.501** ±0.02 (54.1)	0.365 ±0.02	0.569** ±0.02 (55.8)	0.321 ±0.01	0.515** ±0.03 (60.4)	0.347 ±0.02	0.568** ±0.02 (63.6)	0.361 ±0.01	0.589** ±0.02 (63.1)

Values are mean ± SEM of six observations each from tissues pooled from 6 animals

**Values are significant at $p<0.01$

* indicate significance at $p<0.05$

Not significant

Table 6.2 : Changes in Epinephrine Content (μg/g wet wt) in Olfactory Lobe (OL), Cerebellum (CB), Cerebral Cortex (CC) Hippocampus(Hc), Pons medulla (Pm) and Spinal Cord (SPC) of male albino mice on prolonged exposure to Galantamine Hydrobromide. Values in parantheses indicate percent changes from control.

CHRONIC																								
	C	195 DAYS	C	210 DAYS	C	225 DAYS	C	240 DAYS	C	255 DAYS	C	270 DAYS	C	285 DAYS	C	300 DAYS	C	315 DAYS	C	330 DAYS	C	345 DASYS	C	360 DAYS
OL	0.215 ± 0.01	0.335** ± 0.02 (55.8)	0.218 ±0.01	0.335** ±0.01 (53.6)	0.232 ±0.01	0.350** ±0.01 (50.8)	0.243 ±0.02	0.362** ±0.01 (48.9)	0.235 ±0.01	0.345** ±0.01 (46.8)	0.222 ±0.008	0.312** ±0.02 (40.5)	0.271 ±0.02	0.357* ±0.02 (32.1)	0.256 ±0.01	0.332** ±0.01 (29.6)	0.258 ±0.01	0.330** ±0.01 (27.9)	0.275 ±0.01	0.340** ±0.01 (22.8)	0.252 ±0.01	0.301* ±0.004 (19.4)	0.260 ±0.01	0.301** ±0.004 (15.7)
CB	0.421 ±0.01	0.656** ±0.02 (55.8)	0.398 ±0.02	0.635** ±0.02 (59.8)	0.455 ±0.02	0.689** ±0.02 (51.4)	0.415 ±0.02	0.615** ±0.02 (48.1)	0.432 ±0.02	0.635** ±0.03 (46.9)	0.401 ±0.02	0.578** ±0.02 (44.1)	0.435 ±0.02	0.612** ±0.02 (40.6)	0.461 ±0.01	0.632** ±0.02 (37)	0.443 ±0.02	0.601** ±0.02 (35.6)	0.465 ±0.02	0.598** ±0.008 (28.6)	0.473 ±0.01	0.585** ±0.008 (23.6)	0.491 ±0.01	0.595** ±0.02 (21.1)
HC	0.591 ±0.01	0.875** ±0.008 (48)	0.611 ±0.004	0.898** ±0.004 (46.9)	0.596 ±0.01	0.859** ±0.01 (44)	0.626 ±0.01	0.878** ±0.004 (40.2)	0.645 ±0.01	0.883** ±0.02 (36.8)	0.642 ±0.01	0.854** ±0.004 (33)	0.650 ±0.01	0.845** ±0.004 (30)	0.674 ±0.03	0.860** ±0.02 (27.5)	0.700 ±0.004	0.872** ±0.02 (24.5)	0.622 ±0.01	0.765** ±0.01 (22.9)	0.632 ±0.004	0.742** ±0.008 (19.4)	0.626 ±0.008	0.755** ±0.004 (18.5)
CC	0.518 ±0.02	0.800** ±0.02 (54.4)	0.529 ±0.01	0.798** ±0.01 (50.8)	0.528 ±0.01	0.778** ±0.02 (47.3)	0.543 ±0.02	0.785** ±0.02 (44.5)	0.560 ±0.02	0.798** ±0.02 (40.4)	0.535 ±0.02	0.738** ±0.01 (37.9)	0.514 ±0.01	0.698** ±0.01 (35.7)	0.519 ±0.01	0.765** ±0.01 (33.7)	0.553 ±0.01	0.715** ±0.008 (29.2)	0.543 ±0.02	0.698** ±0.02 (28.5)	0.538 ±0.01	0.674** ±0.02 (25.2)	0.568 ±0.02	0.685** ±0.02 (14.5)
PM	0.637 ±0.01	0.981** ±0.003 (54)	0.652 ±0.01	0.941** ±0.02 (44.3)	0.678 ±0.02	0.968** ±0.02 (42.7)	0.701 ±0.02	0.986** ±0.02 (40.6)	0.710 ±0.008	0.958** ±0.02 (34.9)	0.705 ±0.008	0.935** ±0.01 (32.6)	0.721 ±0.02	0.933** ±0.02 (29.4)	0.698 ±0.06	0.885* ±0.02 (26.7)	0.692 ±0.02	0.856** ±0.02 (23.6)	0.716 ±0.01	0.861** ±0.02 (21.2)	0.725 ±0.03	0.862** ±0.02 (18.8)	0.737 ±0.02	0.847* ±0.02 (14.9)
SPC	0.311 ±0.02	0.519** ±0.03 (65.5)	0.348 ±0.02	0.565** ±0.02 (62.3)	0.351 ±0.01	0.564** ±0.03 (60.6)	0.315 ±0.02	0.489** ±0.02 (55.23)	0.336 ±0.02	0.498** ±0.02 (48.2)	0.327 ±0.02	0.465** ±0.02 (42.2)	0.352 ±0.02	0.498** ±0.02 (41.4)	0.368 ±0.02	0.501** ±0.02 (36.1)	0.323 ±0.01	0.415** ±0.02 (28.4)	0.385 ±0.02	0.478** ±0.02 (26.9)	0.341 ±0.02	0.425** ±0.01 (24.6)	0.335 ±0.01	0.404* ±0.02 (20.5)

Values are mean ± SEM of six observations each from tissues pooled from 6 animals

**Values are significant at $p<0.01$

* indicate significance at $p<0.05$

Not significant

Table 7.1 : Changes in Serotonin (5-HT) Content (µg/g wet wt) in Olfactory Lobe (OL), Cerebellum (CB), Cerebral Cortex (CC) Hippocampus(Hc), Pons medulla (Pm) and Spinal Cord (SPC) of male albino mice on prolonged exposure to Galantamine Hydrobromide. Values in parantheses indicate percent changes from control.

	CHRONIC																							
	C	15 DAYS	C	30 DAYS	C	45 DAYS	C	60 DAYS	C	75 DAYS	C	90 DAYS	C	105 DAYS	C	120 DAYS	C	135 DAYS	C	150 DAYS	C	165 DASYS	C	180 DAYS
OL	0.248 ± 0.01	0.305* ± 0.02 (22.98)	0.236 ±0.006	0.295* ±0.02 (25)	0.254 ±0.01	0.333** ±0.01 (31.10)	0.221 ±0.004	0.295* ±0.02 (33.48)	0.218 ±0.008	0.312** ±0.02 (43.11)	0.216 ±0.008	0.315** ±0.01 (45.83)	0.211 ±0.008	0.313** ±0.008 (48.34)	0.211 ±0.008	0.321** ±0.01 (52.13)	0.216 ±0.008	0.339** ±0.01 (56.94)	0.218 ±0.008	0.365** ±0.02 (67.43)	0.207 ±0.008	0.366** ±0.02 (76.81)	0.207 ±0.008	0.354** ±0.01 (71.01)
CB	0.342 ±0.01	0.415* ±0.01 (21.34)	0.322 ±0.01	0.446** ±0.02 (23.75)	0.311 ±0.01	0.422** ±0.02 (35.69)	0.309 ±0.01	0.446** ±0.02 (44.33)	0.317 ±0.01	0.468** ±0.02 (47.63)	0.306 ±0.01	0.452** ±0.02 (47.7)	0.306 ±0.01	0.456** ±0.02 (49.01)	0.311 ±0.01	0.469** ±0.02 (50.80)	0.312 ±0.01	0.478** ±0.02 (53.20)	0.312 ±0.01	0.485** ±0.02 (55.44)	0.318 ±0.01	0.498** ±0.02 (56.60)	0.301 ±0.008	0.495** ±0.02 (64.45)
HC	0.613 ±0.02	0.698* ±0.02 (13.86)	0.608 ±0.02	0.712* ±0.03 (17.10)	0.611 ±0.02	0.735** ±0.02 (20.29)	0.623 ±0.02	0.754** ±0.02 (21.02)	0.578 ±0.02	0.718** ±0.02 (24.22)	0.619 ±0.03	0.778** ±0.02 (25.68)	0.626 ±0.02	0.848** ±0.02 (35.46)	0.616 ±0.03	0.845** ±0.02 (37.17)	0.633 ±0.02	0.896** ±0.02 (41.54)	0.656 ±0.03	0.932** ±0.03 (42.80)	0.615 ±0.03	0.908** ±0.02 (4764)	0.665 ±0.03	0.994** ±0.02 (49.42)
CC	0.533 ±0.02	0.625** ±0.02 (17.26)	0.513 ±0.02	0.658** ±0.01 (28.26)	0.515 ±0.02	0.669** ±0.01 (29.90)	0.541 ±0.02	0.713** ±0.02 (31.79)	0.511 ±0.01	0.698** ±0.02 (36.59)	0.510 ±0.01	0.701** ±0.02 (37.45)	0.544 ±0.02	0.765** ±0.01 (40.62)	0.516 ±0.01	0.745** ±0.01 (44.37)	0.532 ±0.01	0.785** ±0.01 (47.55)	0.520 ±0.01	0.778** ±0.02 (49.61)	0.496 ±0.01	0.795** ±0.01 (60.28)	0.505 ±0.02	0.785** ±0.01 (55.44)
PM	0.536 ±0.01	0.668** ±0.03 (24.62)	0.531 ±0.02	0.668** ±0.03 (25.80)	0.576 ±0.02	0.741** ±0.02 (28.64)	0.551 ±0.01	0.714** ±0.02 (29.58)	0.522 ±0.02	0.685** ±0.02 (31.22)	0.525 ±0.02	0.711** ±0.02 (35.42)	0.506 ±0.02	0.718** ±0.02 (41.89)	0.516 ±0.02	0.748** ±0.01 (44.46)	0.512 ±0.02	0.778** ±0.01 (51.95)	0.515 ±0.02	0.785** ±0.01 (52.42)	0.508 ±0.02	0.779** ±0.01 (53.34)	0.515 ±0.02	0.798** ±0.02 (54.95)
SPC	0.241 ±0.008	0.312** ±0.004 (29.46)	0.254 ±0.01	0.369** ±0.02 (45.27)	0.212 ±0.008	0.314** ±0.02 (48.11)	0.225 ±0.008	0.336** ±0.02 (49.33)	0.217 ±0.01	0.325** ±0.02 (49.76)	0.216 ±0.008	0.324** ±0.02 (50)	0.235 ±0.01	0.358** ±0.02 (52.34)	0.219 ±0.01	0.336** ±0.02 (53.42)	0.206 ±0.008	0.318** ±0.02 (54.36)	0.212 ±0.01	0.358** ±0.02 (68.86)	0.222 ±0.01	0.398** ±0.02 (79.27)	0.201 ±0.008	0.378** ±0.02 (88.05)

Values are mean ± SEM of six observations each from tissues pooled from 6 animals

**Values are significant at $p<0.01$

* Indicate significance at $p<0.05$

Not significant

Table 7.2 : Changes in Serotonin Content (µg/g wet wt) in Olfactory Lobe (OL), Cerebellum (CB), Cerebral Cortex (CC) Hippocampus(Hc), Pons medulla (Pm) and Spinal Cord (SPC) of male albino mice on prolonged exposure to Galantamine Hydrobromide. Values in parantheses indicate percent changes from control.

CHRONIC

	C	195 DAYS	C	210 DAYS	C	225 DAYS	C	240 DAYS	C	255 DAYS	C	270 DAYS
OL	0.219 ± 0.008	0.378** ± 0.02 (72.60)	0.285 ±0.008	0.336** ±0.01 (63.90)	0.218 ±0.004	0.339** ±0.01 (55.50)	0.215 ±0.068	0.328** ±0.01 (52.55)	0.241 ±0.01	0.366** ±0.02 (51.86)	0.208 ±0.008	0.311** ±0.01 (49.51)
CB	0.315 ±0.01	0.503** ±0.02 (59.68)	0.322 ±0.01	0.512** ±0.02 (59)	0.312 ±0.008	0.485** ±0.02 (55.44)	0.322 ±0.01	0.498** ±0.02 (54.65)	0.331 ±0.01	0.502** ±0.02 (51.66)	0.311 ±0.008	0.462** ±0.02 (48.55)
HC	0.598 ±0.02	0.889** ±0.02 (48.66)	0.616 ±0.03	0.906** ±0.02 (47.07)	0.601 ±0.02	0.868** ±0.02 (44.42)	0.633 ±0.03	0.908** ±0.02 (43.44)	0.641 ±0.02	0.923** ±0.04 (43.99)	0.613 ±0.03	0.869** ±0.02 (41.76)
CC	0.516 ±0.01	0.798** ±0.01 (54.65)	0.501 ±0.02	0.765** ±0.02 (52.69)	0.508 ±0.01	0.775** ±0.01 (52.59)	0.520 ±0.01	0.779** ±0.01 (49.80)	0.485 ±0.02	0.715** ±0.02 (47.42)	0.552 ±0.01	0.801** ±0.01 (45.10)
PM	0.501 ±0.01	0.798** ±0.02 (59.28)	0.501 ±0.01	0.789** ±0.02 (57.68)	0.513 ±0.01	0.774** ±0.01 (50.87)	0.508 ±0.01	0.748** ±0.02 (47.24)	0.518 ±0.01	0.752** ±0.02 (45.17)	0.522 ±0.01	0.751** ±0.02 (43.86)
SPC	0.214 ±0.01	0.365** ±0.01 (70.56)	0.215 ±0.008	0.366** ±0.01 (70.23)	0.219 ±0.008	0.345** ±0.01 (62.10)	0.201 ±0.01	0.322** ±0.02 (60.19)	0.207 ±0.01	0.324** ±0.02 (56.52)	0.219 ±0.01	0.336** ±0.02 (53.42)

	C	285 DAYS	C	300 DAYS	C	315 DAYS	C	330 DAYS	C	345 DASYS	C	360 DAYS
OL	0.216 ±0.008	0.322** ±0.01 (49.07)	0.221 ±0.008	0.316** ±0.01 (42.98)	0.209 ±0.004	0.300** ±0.008 (43.54)	0.201 ±0.004	0.277** ±0.01 (37.81)	0.216 ±0.008	0.288** ±0.01 (33.33)	0.207 ±0.008	0.258* ±0.02 (24.63)
CB	0.306 ±0.008	0.445** ±0.02 (45.42)	0.318 ±0.01	0.428** ±0.02 (34.59)	0.309 ±0.01	0.411** ±0.02 (33)	0.316 ±0.01	0.405** ±0.02 (28.16)	0.312 ±0.008	0.399** ±0.02 (27.88)	0.351 ±0.01	0.412* ±0.02 (17.37)
HC	0.626 ±0.02	0.858** ±0.02 (37.06)	0.636 ±0.03	0.854** ±0.02 (34.27)	0.615 ±0.02	0.798** ±0.01 (29.75)	0.578 ±0.02	0.731** ±0.02 (26.47)	0.626 ±0.03	0.784** ±0.01 (25.23)	0.608 ±0.03	0.735* ±0.02 (20.88)
CC	0.532 ±0.02	0.754** ±0.01 (41.72)	0.523 ±0.01	0.714** ±0.03 (36.52)	0.526 ±0.01	0.718** ±0.03 (36.50)	0.532 ±0.01	0.685** ±0.03 (28.75)	0.517 ±0.02	0.654** ±0.01 (26.99)	0.528 ±0.01	0.625* ±0.03 (18.37)
PM	0.542 ±0.01	0.766** ±0.02 (41.32)	0.506 ±0.01	0.712** ±0.01 (40.71)	0.531 ±0.01	0.741** ±0.01 (39.54)	0.516 ±0.01	0.668** ±0.02 (29.45)	0.521 ±0.01	0.663** ±0.02 (27.25)	0.526 ±0.01	0.642** ±0.03 (22.05)
SPC	0.218 ±0.01	0.322** ±0.02 (47.70)	0.261 ±0.01	0.365** ±0.01 (39.84)	0.251 ±0.01	0.333** ±0.02 (31.66)	0.226 ±0.004	0.287* ±0.02 (26.99)	0.254 ±0.01	0.312** ±0.004 (22.83)	0.219 ±0.008	0.266** ±0.01 (21.46)

Values are mean ± SEM of six observations each from tissues pooled from 6 animals

**Values are significant at $p<0.01$

* Indicate significance at $p<0.05$

Not significant

Table 8.1 : Changes in Monoamine Oxidase Content (μg/g wet wt) in Olfactory Lobe (OL), Cerebellum (CB), Cerebral Cortex (CC) Hippocampus(Hc), Pons medulla (Pm) and Spinal Cord (SPC) of male albino mice on prolonged exposure to Galantamine Hydrobromide. Values in parantheses indicate percent changes from control.

	CHRONIC																							
	C	15 DAYS	C	30 DAYS	C	45 DAYS	C	60 DAYS	C	75 DAYS	C	90 DAYS	C	105 DAYS	C	120 DAYS	C	135 DAYS	C	150 DAYS	C	165 DASYS	C	180 DAYS
OL	2.226 ±0.06	1.989* ±0.06 (-10.64)	2.216 ±0.05	1.882* ±0.11 (-15.87)	2.460 ±0.05	2.010** ±0.01 (-18.29)	2.514 ±0.11	2.012** ±0.06 (-19.96)	2.121 ±0.04	1.666** ±0.05 (-21.45)	2.665 ±0.04	2.010** ±0.01 (-24.57)	2.121 ±0.12	1.569** ±0.08 (-26.03)	2.775 ±0.06	2.012** ±0.03 (-27.49)	2.310 ±0.04	1.658** ±0.07 (-28.22)	2.121 ±0.05	1.452** ±0.07 (-31.54)	2.421 ±0.05	1.625** ±0.07 (-32.87)	2.552 ±0.08	1.662** ±0.12 (-34.87)
CB	2.346 ±0.06	2.010* ±0.09 (-14.32)	2.486 ±0.06	2.113** ±0.07 (-15.00)	2.568 ±0.03	2.112** ±0.04 (-17.75)	2.659 ±0.08	2.112** ±0.02 (-20.57)	2.765 ±0.13	2.141** ±0.02 (-22.56)	2.182 ±0.06	1.685** ±0.09 (-22.77)	2.668 ±0.04	2.032** ±0.06 (-23.83)	2.789 ±0.07	2.100** ±0.13 (-24.70)	2.699 ±0.08	2.012** ±0.02 (-25.45)	2.105 ±0.068	1.196** ±0.07 (-43.18)	2.944 ±0.11	1.321** ±0.06 (-55.12)	2.105 ±0.09	1.238** ±0.04 (-41.18)
HC	3.326 ±0.06	2.968** ±0.06 (-10.76)	3.216 ±0.06	2.539** ±0.04 (-21.05)	3.215 ±0.06	2.528** ±0.11 (-21.36)	3.336 ±0.08	2.589** ±0.06 (-22.39)	3.313 ±0.08	2.565** ±0.05 (-22.57)	3.365 ±0.09	2.556** ±0.09 (-24.03)	3.161 ±0.05	2.244** ±0.06 (-29)	3.535 ±0.13	2.369** ±0.08 (-32.98)	3.328 ±0.06	2.122** ±0.06 (-36.23)	3.382 ±0.08	2.115** ±0.10 (-37.46)	3.790 ±0.10	2.135** ±0.05 (-41.80)	3.865 ±0.05	2.119** ±0.05 (-45.27)
CC	3.336 ±0.07	2.964* ±0.11 (-11.19)	3.120 ±0.04	1.556** ±0.08 (-18.07)	3.412 ±0.07	2.556** ±0.06 (-25.08)	3.216 ±0.05	2.141** ±0.03 (-33.42)	3.912 ±0.06	2.557** ±0.11 (-34.63)	3.443 ±0.08	2.106** ±0.07 (-38.83)	3.561 ±0.10	2.013** ±0.05 (-43.47)	3.844 ±0.08	1.996** ±0.05 (-47.57)	3.212 ±0.05	1.553** ±0.07 (-51.65)	3.618 ±0.07	1.552** ±0.09 (-57.10)	3.776 ±0.06	1.345** ±0.18 (-64.38)	3.714 ±0.06	1.225** ±0.06 (-67.01)
PM	3.189 ±0.04	2.669** ±0.06 (-16.30)	3.222 ±0.06	2.665** ±0.08 (-17.28)	3.186 ±0.03	2.552** ±0.07 (-19.89)	3.320 ±0.08	2.622** ±0.06 (-21.02)	3.268 ±0.06	2.441** ±0.08 (-25.30)	3.236 ±0.04	2.356** ±0.06 (27.19)	3.216 ±0.06	2.312** ±0.11 (-28.10)	3.112 ±0.04	2.165** ±0.08 (-30.43)	3.138 ±0.05	2.176** ±0.06 (-30.65)	3.268 ±0.06	2.226** ±0.05 (-31.88)	3.116 ±0.03	2.012** ±0.05 (-35.43)	3.418 ±0.12	2.010** ±0.05 (-41.19)
SPC	2.366 ±0.06	1.125** ±0.07 (-17.90)	2.352 ±0.07	1.212** ±0.06 (-24.07)	2.115 ±0.02	1.116** ±0.04 (-24.71)	2.109 ±0.04	1.140** ±0.05 (-36.95)	2.144 ±0.04	1.236** ±0.05 (-37.55)	2.112 ±0.02	1.232** ±0.02 (-38.10)	2.156 ±0.04	1.326** ±0.04 (-38.25)	2.142 ±0.05	1.478** ±0.07 (-40.97)	2.146 ±0.04	1.552** ±0.07 (-41.26)	2.461 ±0.08	1.872** ±0.12 (-45.90)	2.131 ±0.04	1.689** ±0.08 (-48.67)	2.152 ±0.04	1.785** ±0.07 (-49.86)

Values are mean ± SEM of six observations each from tissues pooled from 6 animals

**Values are significant at $p<0.01$

* Indicate significance at $p<0.05$

Not significant

Table 8.2 : Changes in Monoamine Oxidase Content (μg/g wet wt) in Olfactory Lobe (OL), Cerebellum (CB), Cerebral Cortex (CC) Hippocampus(Hc), Pons medulla (Pm) and Spinal Cord (SPC) of male albino mice on prolonged exposure to Galantamine Hydrobromide. Values in parantheses indicate percent changes from control.

CHRONIC																								
	C	195 DAYS	C	210 DAYS	C	225 DAYS	C	240 DAYS	C	255 DAYS	C	270 DAYS	C	285 DAYS	C	300 DAYS	C	315 DAYS	C	330 DAYS	C	345 DASYS	C	360 DAYS
OL	2.552 ± 0.07	1.252** ± 0.03 (-50.94)	2.669 ±0.07	1.658** ±0.04 (-38.56)	2.158 ±0.04	1.359** ±0.07 (-37.02)	2.332 ±0.05	1.625** ±0.06 (-30.31)	2.223 ±0.08	1.565** ±0.08 (-29.59)	2.366 ±0.06	1.658** ±0.05 (-29.52)	2.326 ±0.06	1.658** ±0.10 (-28.71)	2.165 ±0.03	1.559** ±0.07 (-27.99)	2.336 ±0.06	1.856** ±0.05 (-20.54)	2.685 ±0.08	2.221** ±0.05 (-17.28)	2.525 ±0.06	2.104** ±0.04 (-16.67)	2.141 ±0.03	1.912** ±0.06 (-10.69)
CB	2.336 ±0.07	1.559** ±0.07 (-33.26)	2.464 ±0.07	1.659** ±0.09 (-32.67)	2.958 ±0.02	2.100** ±0.05 (-29)	2.125 ±0.08	1.526** ±0.13 (-28.18)	2.895 ±0.13	2.116** ±0.04 (-26.90)	2.856 ±0.10	2.113** ±0.02 (-26.01)	2.641 ±0.09	2.030** ±0.08 (-23.13)	2.662 ±0.08	2.060* ±0.20 (-22.61)	2.659 ±0.09	2.131** ±0.02 (-19.85)	2.451 ±0.06	2.015** ±0.05 (-17.78)	2.451 ±0.05	2.015** ±0.08 (-14.11)	2.433 ±0.04	2.121** ±0.04 (-12.82)
HC	3.512 ±0.05	2.000** ±0.06 (-43.05)	3.441 ±0.11	2.021** ±0.11 (-41.26)	3.261 ±0.04	2.012** ±0.01 (-38.30)	3.622 ±0.10	2.325** ±0.06 (-35.80)	3.861 ±0.08	2.560** ±0.09 (-33.69)	3.116 ±0.02	2.148** ±0.04 (-31.06)	3.121 ±0.06	2.156** ±0.05 (-30.91)	3.562 ±0.07	2.518** ±0.08 (-29.30)	3.372 ±0.09	2.659** ±0.04 (-21.14)	3.911 ±0.12	3.201** ±0.03 (-18.15)	3.586 ±0.08	3.011** ±0.11 (-16.03)	3.169 ±0.02	2.698** ±0.06 (-14.86)
CC	3.843 ±0.03	1.110** ±0.03 (-71.18)	3.311 ±0.07	1.125** ±0.03 (-66.02)	3.336 ±0.10	1.212** ±0.05 (-63.66)	3.561 ±0.09	1.556** ±0.05 (-56.30)	3.965 ±0.07	2.312** ±0.04 (-41.68)	3.545 ±0.07	2.331** ±0.08 (-34.24)	3.661 ±0.08	2.458** ±0.10 (-32.85)	3.120 ±0.04	2.116** ±0.04 (-32.17)	3.422 ±0.09	2.565** ±0.04 (-25.04)	3.659 ±0.06	2.996** ±0.08 (-18.11)	3.221 ±0.08	2.663** ±0.07 (-17.32)	3.212 ±0.04	2.885** ±0.08 (-10.18)
PM	3.416 ±0.08	2.110** ±0.05 (-38.23)	3.156 ±0.04	1.986** ±0.08 (-37.07)	3.510 ±0.13	2.216** ±0.15 (-36.86)	3.212 ±0.04	2.116** ±0.04 (-35.21)	3.321 ±0.05	2.156** ±0.03 (-35.07)	3.311 ±0.04	2.115** ±0.04 (-36.12)	3.332 ±0.05	2.169** ±0.05 (-34.90)	3.330 ±0.10	2.180** ±0.08 (-34.53)	3.189 ±0.06	2.114** ±0.14 (-33.70)	3.116 ±0.04	2.156** ±0.09 (-30.80)	3.228 ±0.05	2.316** ±0.06 (-28.25)	3.165 ±0.04	2.331** ±0.07 (-26.35)
SPC	2.313 ±0.08	1.890* ±0.13 (-52.45)	2.181 ±0.08	1.656** ±0.11 (-48.46)	2.169 ±0.07	1.633** ±0.08 (-47.23)	2.162 ±0.18	1.363** ±0.08 (-45.94)	2.122 ±0.03	1.315** ±0.05 (-42.35)	2.136 ±0.01	1.322** ±0.04 (-41.66)	2.010 ±0.06	1.241** ±0.05 (-38.49)	2.121 ±0.02	1.252** ±0.03 (-30.99)	2.113 ±0.02	1.241** ±0.06 (-27.67)	2.322 ±0.09	1.256** ±0.06 (-23.93)	2.231 ±0.04	1.165** ±0.10 (-28.74)	2.216 ±0.05	1.115** ±0.06 (-17.05)

Values are mean ± SEM of six observations each from tissues pooled from 6 animals

** Values are significant at $p<0.01$

* Indicate significance at $p<0.05$

Not significant

Figs. 3,4&5 : Graphical representation of percent changes in the content of Dopamine (In vivo) in Olfactory Lobe (OL), Hippocampus (Hc) and Cerebral Cortex (CC) of experimental mice brain following prolonged exposure to galantamine hydrobromide against the control

Fig. 3

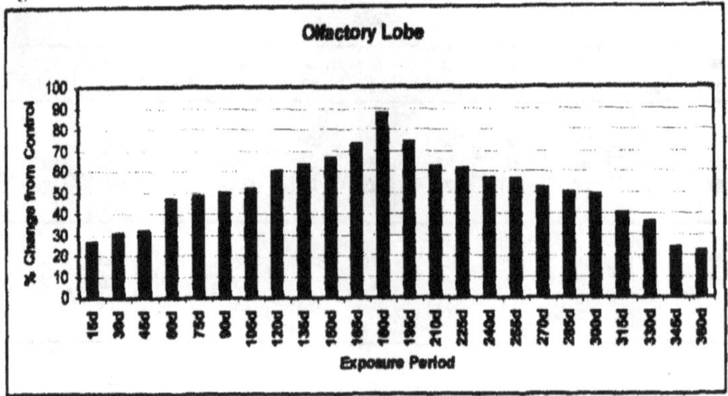

Fig. 4

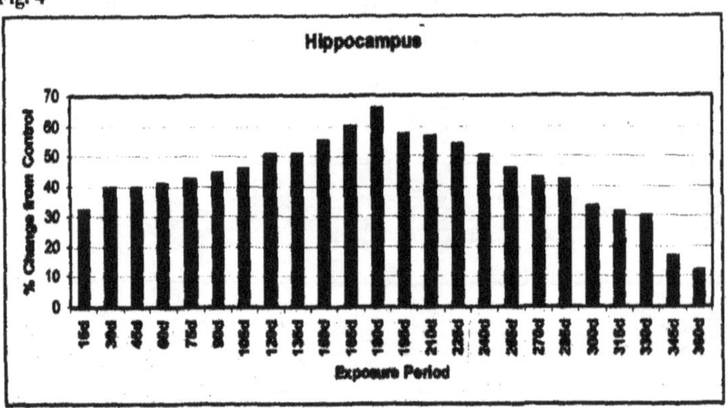

Fig. 5

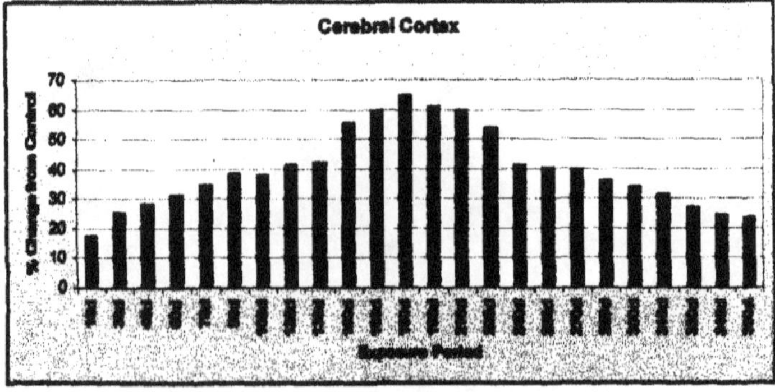

Figs. 6,7&8 : Graphical representation of percent changes in the content of Dopamine (In vivo) in Cerebellum (Cb), Ponsmedulla (Pm), Spinal cord (Spc) of experimental mice brain following prolonged exposure to galantamine hydrobromide against control.

Fig. 6

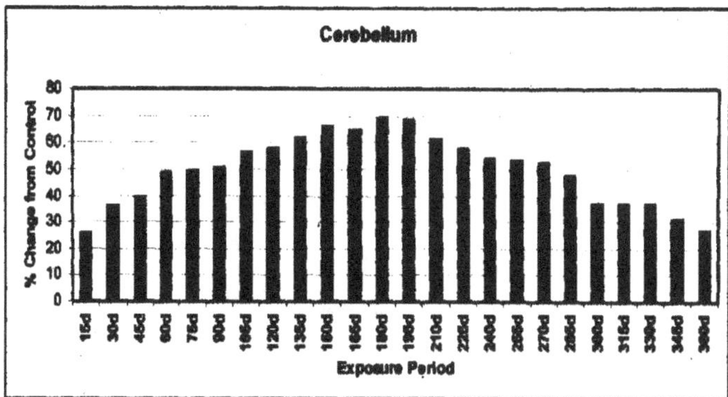

Fig. 7

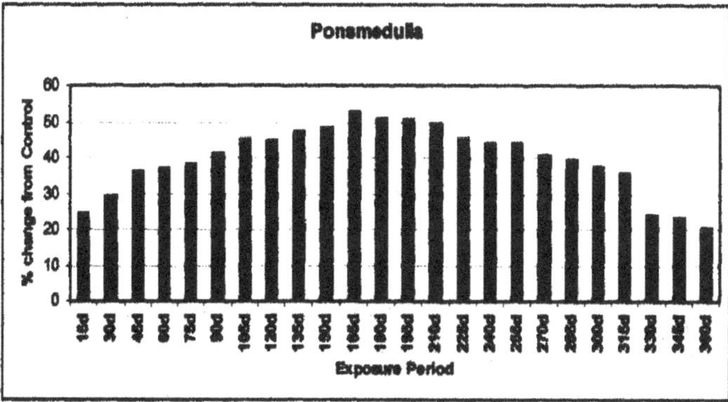

Fig. 8

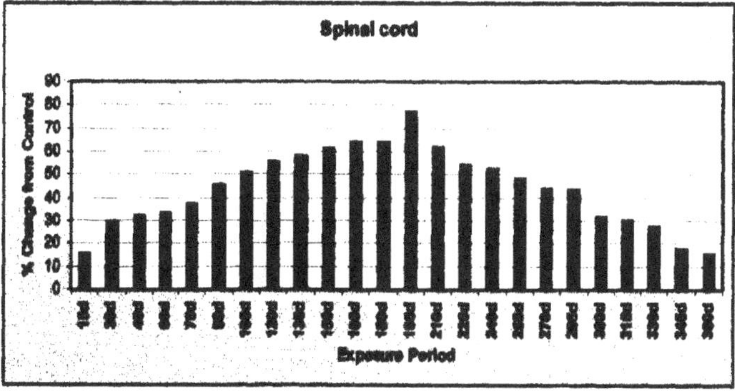

Figs. 9,10&11. : Graphical representation of percent changes in the content of Norepinephrine (In vivo) in Olfactory Lobe (OL), Hippocampus (Hc), Cerebral Cortex (CC) of experimental mice brain following prolonged exposure to galantamine hydrobromide against the control.

Fig. 9

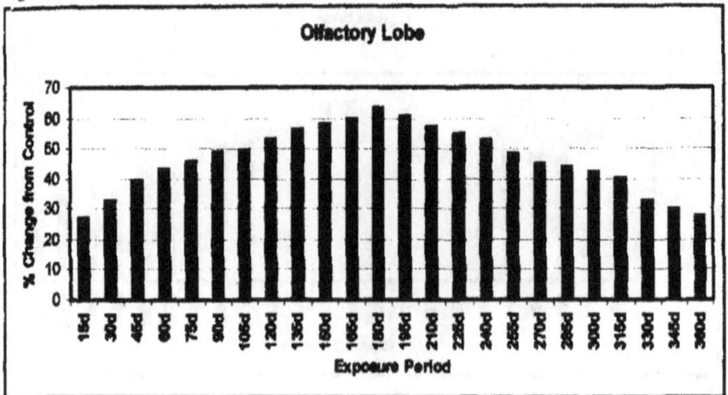

Fig. 10

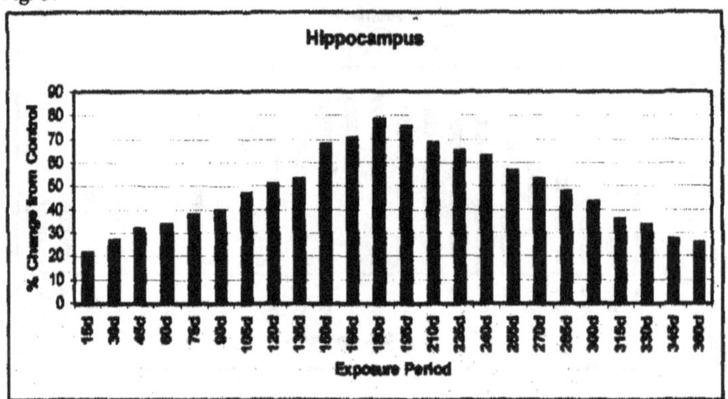

Fig. 11

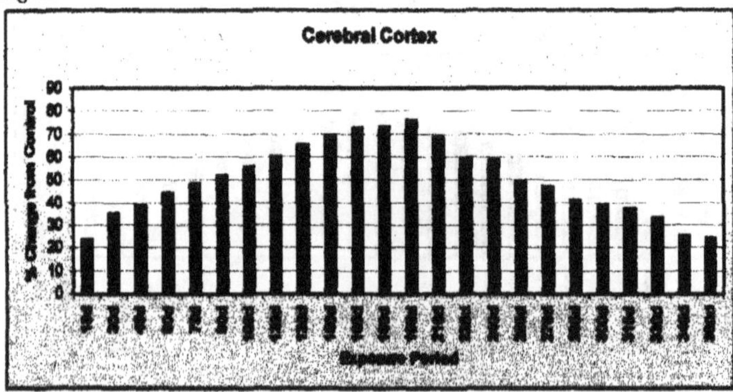

Figs. 12,13&14 : Graphical representation of percent changes in the content of Norepinephrine (In vivo) in Cerebellum (Cb), Ponsmedulla (Pm), Spinal cord (Spc) of experimental mice brain following prolonged exposure to galantamine hydrobromide against the control.

Fig. 12

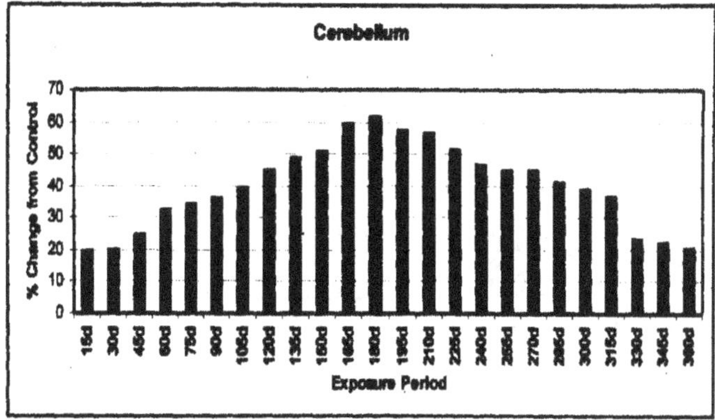

Fig. 13

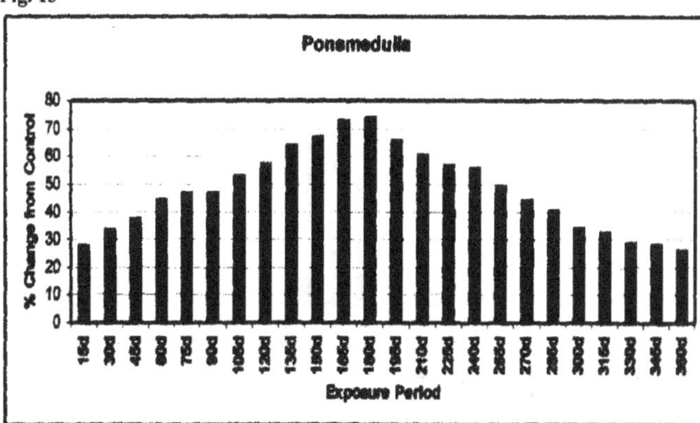

Fig. 14

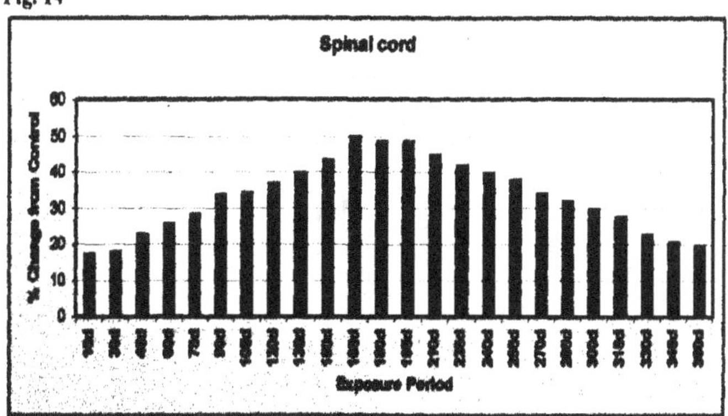

Figs. 15,16&17 : Graphical representation of percent changes in the content of Epinephrine (In vivo) in Olfactory Lobe (OL), Hippocampus (Hc), Cerebral Cortex (CC) of experimental mice brain following prolonged exposure to galantamine hydrobromide against the control.

Fig. 15

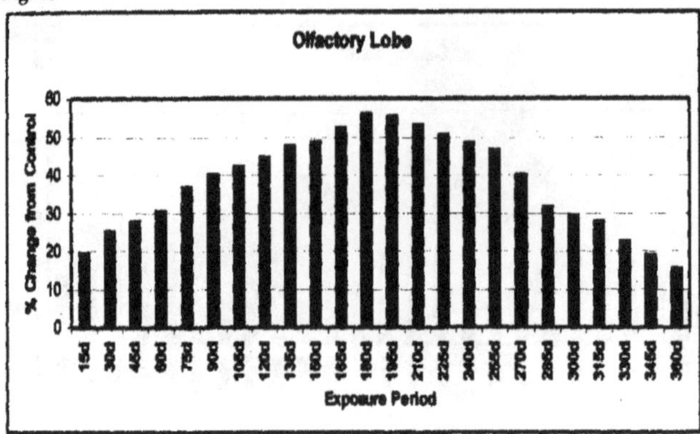

Fig. 16

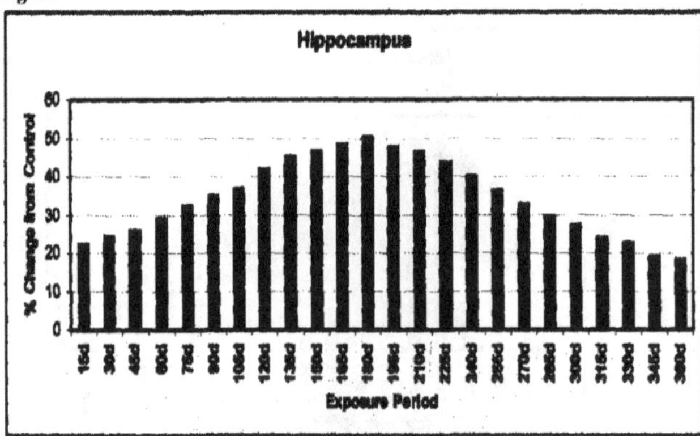

Fig. 17

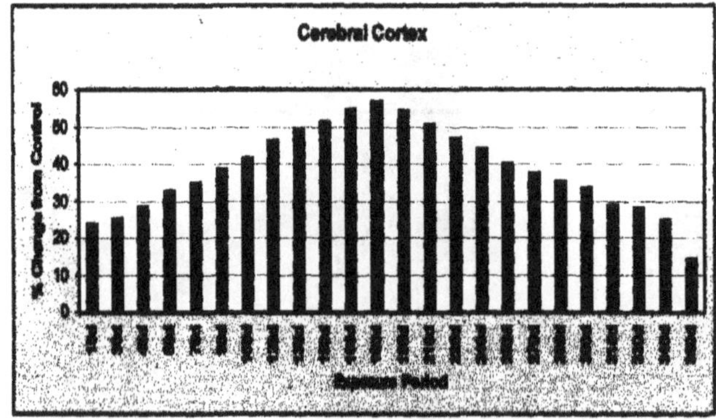

Figs. 18,19&20 : Graphical representation of percent changes in the content of Epinephrine (In vivo) in Cerebellum (Cb), Ponsmedulla (Pm), Spinal cord (Spc) of experimental mice brain following prolonged exposure to galantamine hydrobromide against control.

Fig. 18

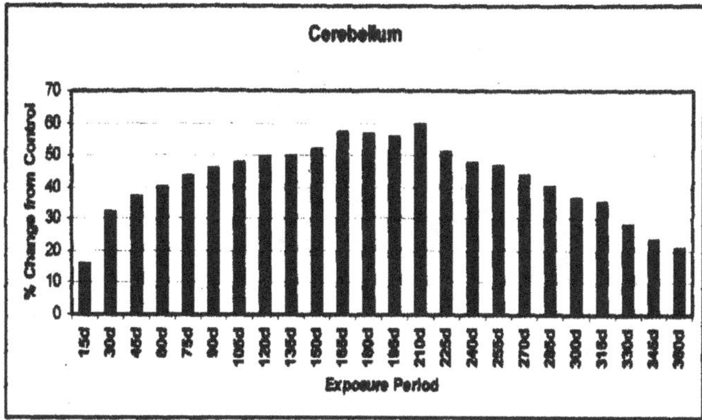

Fig. 19

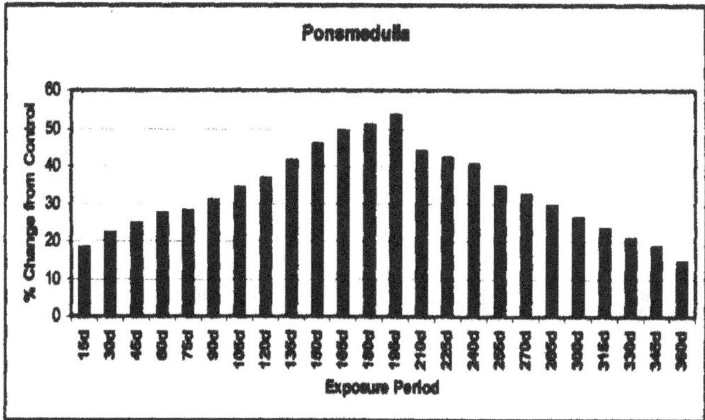

Fig. 20

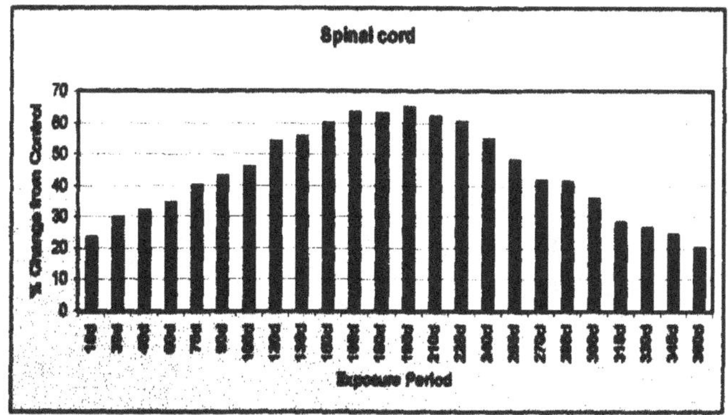

Figs. 21,22&23 : Graphical representation of percent changes in the content of Serotonin (In vivo) in Olfactory Lobe (OL), Hippocampus (Hc), Cerebral Cortex (CC) of experimental mice brain following prolonged exposure to galantamine hydrobromide against the control.

Fig. 21

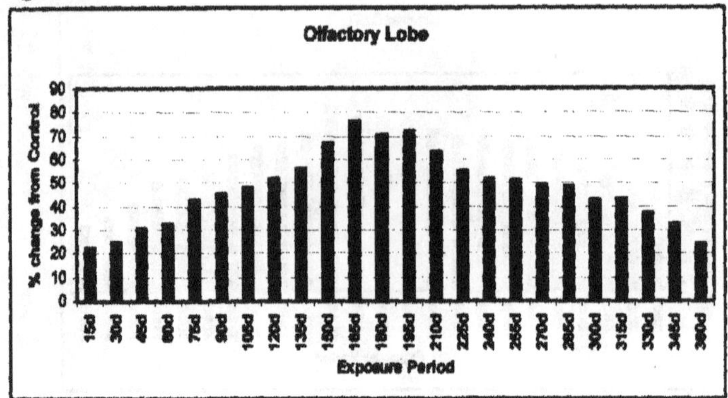

Fig. 22

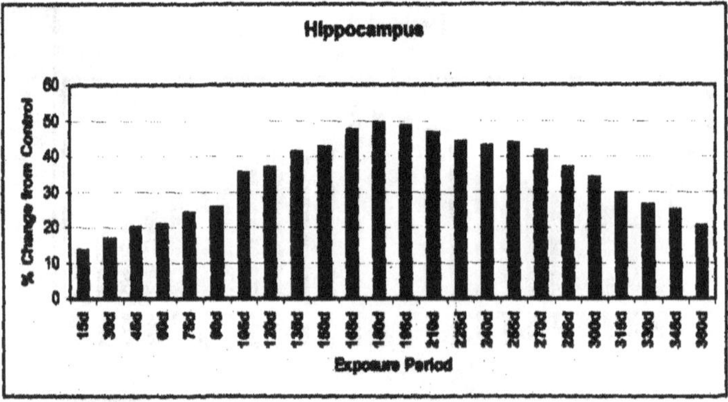

Fig. 23

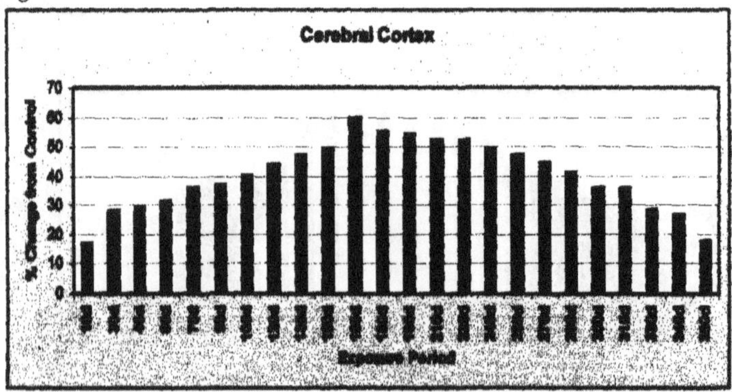

Figs. 24,25&26 : Graphical representation of percent changes in the content of Seotonin (In vivo) in Cerebellum (Cb), Ponsmedulla (Pm), Spinal cord (Spc) of experimental mice brain following prolonged exposure to galantamine hydrobromide against the control.

Fig. 24

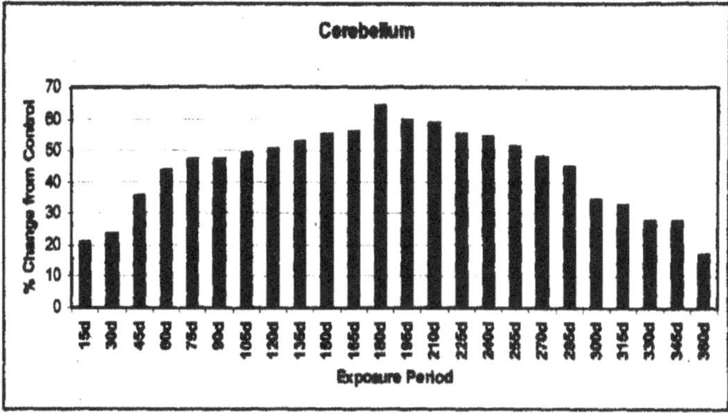

Fig. 25

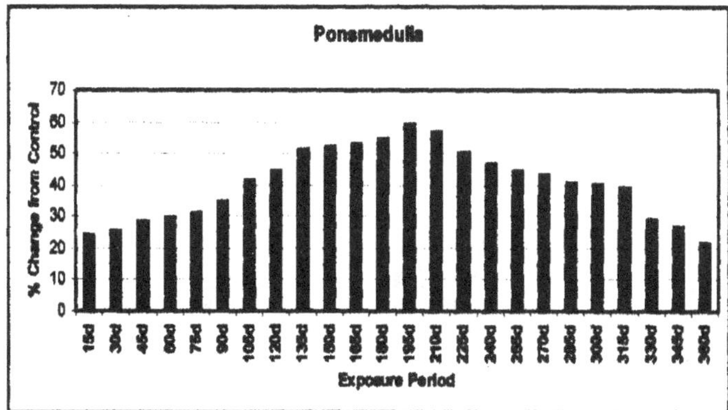

Fig. 26

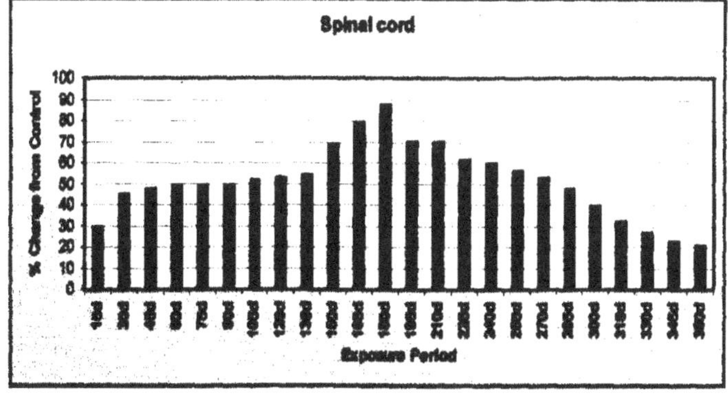

Figs. 27,28&29 : Graphical representation of percent changes in the content of Monoamine oxidase (In vivo) in Olfactory Lobe (OL), Hippocampus (Hc), Cerebral Cortex (CC) of experimental mice brain following prolonged exposure to galantamine hydrobromide against the control.

Fig. 27

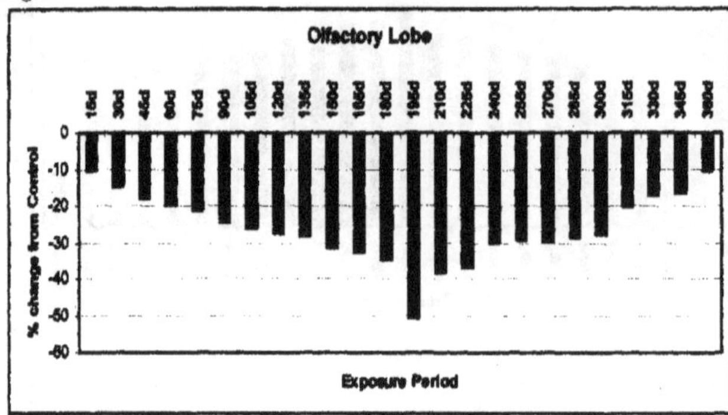

Fig. 28

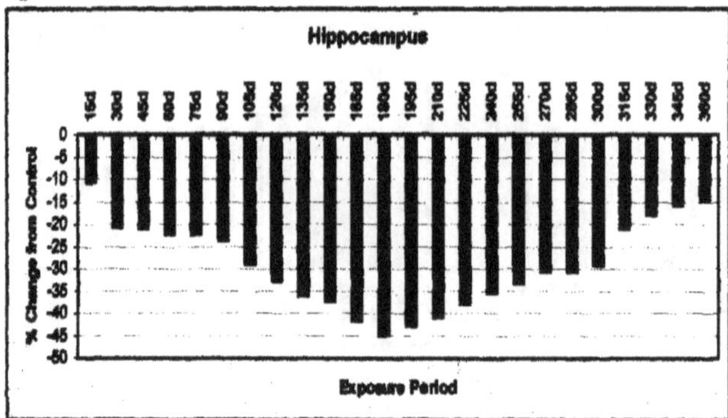

Fig. 29

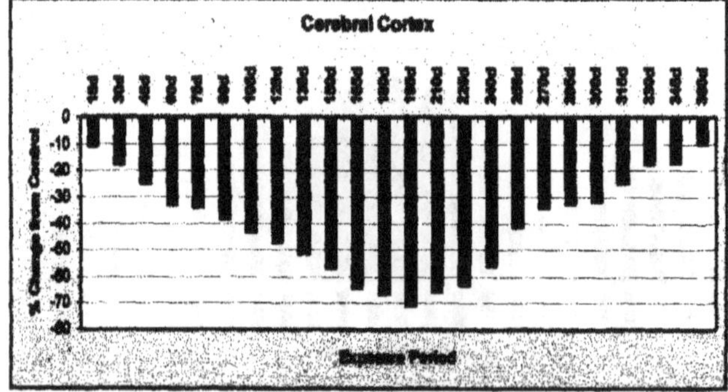

Figs. 30,31&32 : Graphical representation of percent changes in the content of Monoamine oxidase (In vivo) in Cerebellum (Cb), Ponsmedulla (Pm), Spinal cord (Spc) of experimental mice brain following prolonged exposure to galantamine hydrobromide against the control.

Fig. 30

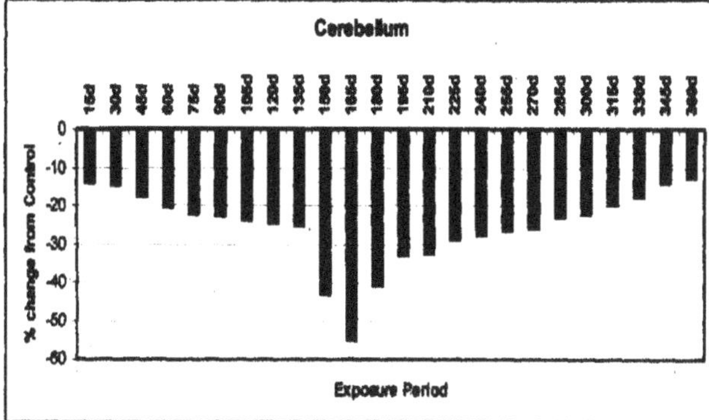

Fig. 31

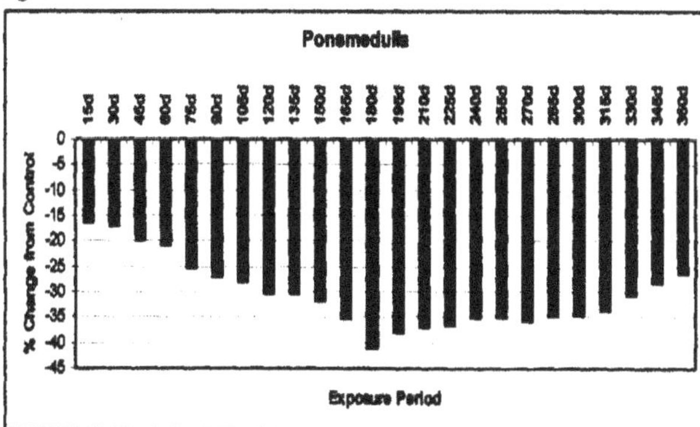

Fig. 32

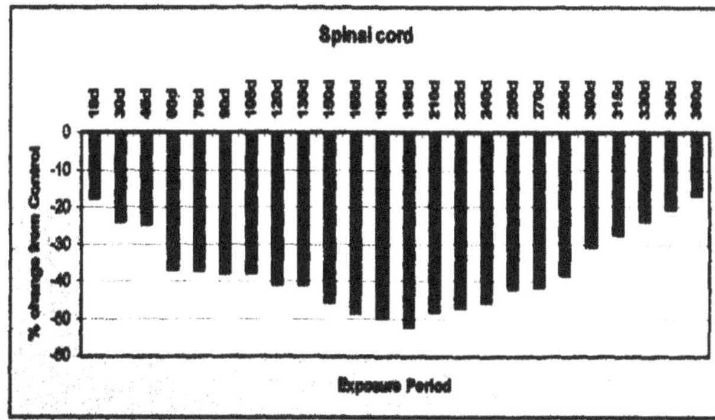

CHAPTER – III

ATPase SYSTEM

Energy for metabolic processes will be derived from hydrolysis of phosphate bonds of ATP by complex enzyme systems involving ATPases (**Christopher, 1979**). Membrane localization is the key to physiological function of ATPases, which is coupled with pumping of cations across the membrane from inside to outside of the cell (**Schwartz *et al.*, 1975**). This process is widely used in all known forms of life. The ATP is hydrolysed by enzymes known as phosphatases (**Lehninger, 2004**). ATPases are insoluble proteins but integral hydrophobic protein components of membranes (**Boyer, 1976**). Some such enzymes are integral membrane proteins and move solutes across the membrane, typically against their concentration gradient. These are called transmembrane ATPases. The energy derived through the consumption of macromolecules, carbohydrates, proteins and lipids is stored as adenosine triphosphate (ATP) and its derivatives in bio system (**Lehninger, 1984**).

The distribution of ATPases is widespread with Na^+, K^+- ATPases mostly present in all nerve cells, Ca^{+2}- ATPases mostly present in muscle cells and Mg^{+2}- ATPases in almost all cells (**Harper *et al.*, 1977**). ATPases have requirement for Mg^{+2}, Ca^{+2}, Na+, K^+ ions for their activity and involve in the cleavage of ATP to ADP/AMP and inorganic phosphate with energy release. Further, the ATPases begin an integrated enzyme of mitochondria, any damage to the mitochondria ultimately alters its activities which would interfere with conversion of oxidative energy to phosphorylated energy. The reaction takes place in the presence of Mg^{+2} and has an absolute requirement for both Na^+ and K^+ (**Abrams *et al.*, 1972**). ATPases are associated with neurotransmitter release, cardiac contractility and other cellular functions such as cell volume regulation and membrane potential (**Satyavathi and Prabhakara rao, 2002**).

Functions

Transmembrane ATPases are important for the import of metabolites necessary for cell metabolism and export toxins, wastes and solutes that can hinder cellular processes. The best example is the sodium-potassium enhancer (or Na^+/K^+- ATPase), which establishes the ionic concentration balance that maintains the cell potential. Another example is the hydrogen potassium ATPase (H^+/K^+- ATPases or gastric proton pump) that acidifies the contents of the stomach.

Besides exchangers, other categories of transmembrane ATPases include co-transporters and pumps (however, some exchanges are also pumps). Some of these, like the Na^{+}/K^{+}- ATPases, cause a net flow of charge, but others do not. These are called "electrogenic" and "nonelectrogenic", transporters respectively.

Mechanism

The coupling between ATP hydrolysis and transport is more or less a strict chemical reaction, in which a fixed number of solute molecules are transported for each ATP molecule that is hydrolyzed; for example, $3Na^{+}$ ions out of the cell and $2K^{+}$ ions inward per ATP hydrolyzed, for the Na^{+}/K^{+} exchanger.

Transmembrane ATPase harnesses the chemical potential energy of ATP, because they perform mechanical work; they transport solutes in a direction opposite to their thermodynamically preferred direction of movement - that is, from the side of the membrane where they are in low concentration to the side where they are in high concentration. This process is considered active transport. For example, the blocking of the vesicular H^{+}- ATPases would increase the P^{H} of the cytoplasm.

Transmembrane ATP Synthases

The ATPsynthase of mitochondria and chloroplasts is an anabolic enzyme that harnesses the energy of a transmembrane proton gradient as an energy source for adding an inorganic phosphate group to a molecule of adenosine diphosphate (ADP) to form a molecule of adenosine triphosphate (ATP).

This enzyme works when a proton moves down the concentration gradient, giving the enzyme a spinning motion. This unique spinning motion bonds ADP and P together to create ATP. ATP synthase can also function in reverse, that is, use energy released by ATP hydrolysis to pump protons against their thermodynamic gradient.

ATPases have been classified based on the requirement of specific cations such as Na^{+}/K^{+}- ATPases, Mg^{+2}- ATPase and Ca^{+2}- ATPase.

Na^{+}/K^{+}- ATPase

Na^{+}/K^{+}- ATPases, mostly present in nerve cells. Na^{+}/K^{+}- ATPases catalyse the conversion of ATP to ADP and Pi. This reaction takes place in the presence of Mg^{+2} and has an absolute requirement for both Na^{+} and K^{+}. Na^{+}/K^{+}- ATPases which are predominantly present in nerve cells **(Matsumura *et al.*, 1969)**. Stimulating membrane bound enzymes are shown to be involved in the active transport, across the cell membrane **(Gylman and Karlish, 1975).**

Active transport of ions Na^+ and K^+ couples with membrane ATPase undergoing phosphorylation and dephosphorylation which in turn supplies energy for the transport of ions against concentration gradient. Sodium potassium balance is maintained by the energy dependent active transport of Na^+ out of the cells and K^+ into the cells promoted by Na^+, K^+ pump which catalyses the transport of three Na^+ ions outward and two K^+ ions inward. The process is coupled by hydrolysis of ATP molecule. Hence sodium potassium ATPases is a biochemical expression of the active transport of Na^+-K^+ ions in cells **(Skou, 1965)** of various cellular process, for which pump is vital in the maintenance of electro chemical gradient across the cell membranes, particularly in nerve and muscle cells **(Thomas, 1972; Barker, 1972).**

The Na^+/K^+- ATPase is not only responsible for a symmetric distribution of Na^+ and K^+ ions across the cell membrane, but also in conjugation with the other transport proteins mediates the bulk movement of ions and fluid in a variety of tissues. The highest activity of Na^+, K^+- ATPse is observed in cerebral cortex which is known to be the region of learning and intelligence **(Prosser, 1973).** The highest activity of the Na^+/K^+- ATPase enzyme is an indication of the active regulation of the ionic pump coupled to the hydrolysis of the ATP to maintain the metabolic status of the tissue **(Shafeek, 2001).**

Purified Na^+/K^+- ATPase contain the larger polypeptide often called the α- sub unit, Molecular Weight = 85,000 to 1,20,000 and a small glycopeptide β- sub unit, Molecular Weight = 42,000 to 60, 000. The α- chain contains the site for phosphorylation **(Kyte, 1971)** and the main ouabain binding site **(Forbush *et al.*, 1978).** The enzyme contains about 36 sulfydril groups per molecule **(Schoot *et al.*, 1978).** The ouabain sensitive Na^+/K^+- ATPase and ouabain insensitive Mg^{+2}- ATPase are membrane enzymes ubiquities in animal cells including neurons but especially abundant in Synaptic Plasma Membranes (SPM) **(Prosser, 1973).**

Na^+/K^+- ATPase is the enzymatic expression of the principal active cation transport system in eukaryotic cells. It is ubiquitous among animal species but its activity ranged widely. Highest activity has been recorded in excitatory and secretary tissues like brain, electric cell electroplax, kidney outer medulla and salt glands of marine birds and fishes (**Bonting, 1970**). The specific activity of Na^+/K^+- ATPase was high in the brain when compared to other tissues.

The ionic gradient produced by this enzyme is coupled to physiological functions such as cell proliferation, volume regulation, maintenance of the electrogenic potential required for the function of excitable tissues, i.e. muscle and nerves and secondary active transport (**Vasilets and Schwartz, 1973; Boldyrev,1993; Basavappa *et al.*, 1998; Rodriguez de Lores Amaiz and Pena, 1995**). Malfunction of Na^+/K^+- ATPase enzyme has been associated with neuronal hyper-excitability, cellular depolarization and swelling (**Lees, 1991**). In numerous tissues, the activities of Na^+/K^+- ATPase may be influenced by different endogenous modulators (**Rodrigez and Pena, 1995; Balzan *et al.*, 2000**) and exogenous factors including certain divalent metals and organic compounds of toxicological interest (**Vasic *et al.*, 2002**) as well as some drugs.

Mg^{+2}- ATPase

It has a unique role in energy synthesis and is localized in mitochondria of all types of cells. Metabolic and structural damage to the mitochondrial membrane results in profound disturbances in the redox system which inturn affect the energy metabolism as well as neuronal transport and other vital processes (**Benzi *et al.*, 1978**). Among different ATPases, Mg^{+2} – ATPase alone constitutes 40-60% of the total ATPases (**Robinson, 1983**). In most cases, Mg^{+2}- ATPase is taken as an index of general ATPase activity because of its abundant distribution and dual localization in mitochondria and cytosol (**Lehninger, 1978**).

Mg^{+2}- ATPase, the major ATPase of the plasma membrane (**Vajreswari and Narayanareddy, 1992**) is responsible for the aminophospholipid translocase activity of the plasma membrane (**Aluland *et al.*, 1994; Bartosz *et al.*, 1994; Vermeuten *et al.*, 1995**) which could affect the activity of the membrane Ca^{+2}- ATPase (**Strittmatter *et al.*, 1979**). Mg^{+2}- ATPase is sensitive to membrane fatty acid composition (**Zimmerman and Daleke, 1993**) which is related to the activity of intracellular calcium – stimulated fatty acid synthase, suggesting a possible synergistic relationship between Mg^{+2}- ATPase and Ca^{+2}- ATPase.

Ca^{+2}- ATPase

Ca^{+2}- ATPase is one of the key enzymes involved in ATP hydrolysis, plays an important role in maintaining a constant intracellular calcium concentration and stimulation of mitochondrial respiration by calcium in the media (**Ohashi *et al.*, 1970; Carafoli and Lehninger, 1978**). The hydrolysis of ATP by Ca^{+2} stimulated ATPase proceeds through a series of partial reactions involving the formation of phosphorylated intermediate (**Garrahan and Rega, 1998**). The Ca^{+2} concentration in the nerve cell is a key factor in the control of neurotransmitter release. The arrival of action potentials at the nerve terminal elicits an increase in the Ca^{+2} concentrations that triggers the neurotransmitter exocytosis. Removal of Ca^{+2} from the nerve terminal is then required to terminate the process neurotransmitter release. Nerve terminals have different ion transport systems that contribute to the regulation of the low intracellular free Ca^{+2} concentrations. The control of intracellular free calcium concentration is crucial for the maintenance of normal cell function and is regulated through the separation of several mechanisms, including of ATP- driven calcium pump (**Lynch and Cheung, 1979**).

Changes in free cytosolic calcium concentration play a vital role in the action of certain hormones and other stimuli on cell metabolism (**Charest *et al.*, 1983; Joseph and Wiliamson, 1983**). The neuronal plasma membrane contains mainly a Ca^{+2}-ATPase that pumps Ca^{+2} to the extracellular medium and a Na^{+}/Ca^{+2} exchanger antiporter that can affect the intracellular Ca^{+2} concentration by modifying the Na+ gradient across the membrane. Within the synaptic terminal, Ca^{+2} can also be removed from the cytosol to the Endoplasmic Reticulum by a Ca^{+2}- ATPase or to the mitochondria by a Ca^{+2} uniporter (**Carafoli and Crampton, 1978**).

Ca^{+2}- ATPase is an important pump that extrudes calcium out of cells (**Carafoli, 1987**) and its normal function requires adequate ATP (**Carafoli, 1987**). The enzyme is also referred to as the calcium pump ATPase owing to its well demonstrated link with Ca^{+2} transport (**Schatzmann and Vincenzi, 1969**). Ca^{+2}- ATPase in different organs generally requires Mg^{+2} for its activation (**Carafoli and Crampton, 1978**). However Ca^{+2}- ATPase without a requirement for Mg^{+2} has also been reported (**Vijayasarathi *et al.*, 1980; Gupta and Venkita Subramanian, 1983**).

In view of the due importance of ATPases in many physiological functions of the nervous system, in the present study an attempt has also been made to determine the impact of galantamine hydrobromide on the ATPase system.

RESULTS

ATPase SYSTEM

Different ATPases such as Na^+, K^+, Mg^{+2}, Ca^{+2} – ATPase activities were estimated in different brain areas of control and galantamine hydrobromide administered mice on selected days as in the case of biogenic amines.

Na^+, K^+ - ATPase activity (Tables. 9.1 & 9.2 Figs. 33 to 38)

Eventhough, the Na^+, K^+ - ATPase activity in general in all areas of mice brain was inhibited under prolonged exposure to galantamine hydrobromide. Between 30th to 180th day, the effect of galantamine hydrobromide on different regions of mice brain was different as shown below.

30th day Spc > CC > Cb > Pm > Hc > OL
(-28.31%) (-22.33%) (-21.44%) (-21.27%) (-19.30%) (-19.02%)

60th day Pm > Spc > OL > Cb > CC > Hc
(-35.12%) (-35.09%) (-32.94%) (-30.55%) (-30.06%) (-27.71%)

90th day OL > Pm > Spc > CC > Hc > Cb
(-39.53%) (-38.42%) (-36.83%) (-36.70%) (-34.27%) (-33.87%)

120th day Spc > OL > Pm > CC > Cb > Hc
(-49.23%) (-46.37%) (-46.02%) (-45.06%) (-43.88%) (-38.52%)

150th day Cb > CC > OL > Spc > Pm > Hc
(-53.45%) (-53.08%) (-52.89%) (-52.06%) (-50.55%) (-47.76%)

180th day Pm > Spc > Cb > OL > CC > Hc
(-63.76%) (-57.80%) (-57.42%) (-57.32%) (-55.75%) (-55.55%)

During subsequent periods, **maximum decrease** in Na^+, K^+ - ATPase activity in different regions was as follows. Ponsmedulla (63.76%) on 180th day; Spinal cord (59.24%) on 165th day; Cerebellum (57.42%) on 180th day; Olfactory Lobe (57.32%) on 180th day; Hippocampus (55.55%) on 180th day; Cerebral Cortex (55.36%) on 195th and 240th day.

Between 210th day and 360th day the depleting trend in of Na^+, K^+ - ATPase activity showed gradual improvement in all the regions of mice brain as follows.

210th day Pm > Spc > Cb > OL > CC > Hc
(-57.58%) (-54.72%) (-53.51%) (-53.27%) (-53.10%) (-47.26%)

240th day CC > Pm > Spc > OL > Cb > Hc
(-55.36%) (-52.10%) (-48.26%) (-46.14%) (-45.35%) (-38.22%)

270th day Pm > CC > Spc > Cb > OL > Hc
(-47.44%) (-44.27%) (-44.24%) (-43.37%) (-39.02%) (-33.89%)

300th day Pm > CC > Spc > OL > Cb > Hc
(-42.43%) (-40.62%) (-36.03%) (-34.96%) (-33.62%) (-33.17%)

330th day Pm > Spc > Hc > Cb > CC > OL
(-36.80%) (-26.93%) (-26.79%) (-26.31%) (-25.98%) (-18.56%)

360th day Pm > Hc > Spc > CC > Cb > OL
(-26.03%) (-19.02%) (-17.78%) (-12.73%) (-12.29%) (-8.18%)

Mg^{+2} – ATPase activity (Tables. 10.1 & 10.2 Figs. 39 to 44)

Similar to Na^{+}, K^{+} - ATPase, Mg^{+2} – ATPase activity also recorded significant inhibition in different brain areas of albino mice treated with galantamine hydrobromide. From 30th day upto 180th day the extent of inhibition in Mg^{+2} – ATPase activity was recorded in the following order.

30th day Spc > CC > Cb > Pm > Hc > OL
(-29.25%) (-22.62%) (-22.35%) (-21.44%) (-20.21%) (-19.83%)

60th day Pm > Spc > OL > Cb > CC > Hc
(-36.20%) (-36.12%) (-34.31%) (-31.23%) (-30.94%) (-28.97%)

90th day OL > Pm > CC > Spc > Cb > Hc
(-41.35%) (-40.10%) (-37.85%) (-36.76%) (-35.45%) (-35.16%)

120th day Spc > OL > Pm > CC > Cb > Hc
(-50.56%) (-49.84%) (-48.43%) (-46.86%) (-45.72%) (-39.95%)

150th day OL > Cb > CC > Spc > Pm > Hc
(-55.54%) (-55.32%) (-54.23%) (-53.82%) (-52.41%) (-49.83%)

Maximum depletion on Mg^{+2} – ATPase activity was observed in different regions of mice brain on 180th day except in Ponsmedulla and Cerebellum.

180th day Pm > Spc > HC > OL > Hc > Cb
(-65.73%) (-61.62%) (-60.96%) (-60.24%) (-58.36%) (-58.06%)

From 210th day onwards upto 360th day, Mg^{+2} – ATPase activity exhibited a reversal trend as observed below.

210th day Pm > CC > Spc > Cb > OL > Hc
(-59.10%) (-56.96%) (-56.53%) (-55.86%) (-55.68%) (-49.01%)

240th day Pm > CC > Spc > Cb > OL > Hc
(-61.84%) (-56.91%) (-50.31%) (-48.22%) (-48.01%) (-47.11%)

270th day Pm > Spc > CC > Cb > OL > HC
(-48.66%) (-46.04%) (-45.97%) (-44.85%) (-41.69%) (-37.32%)

300th day Pm > CC > Spc > OL > Cb > HC
(-41.72%) (-39.98%) (-37.61%) (-36.54%) (-34.91%) (-34.55%)

330th day Pm > Spc > Hc > Cb > CC > OL
(-29.80%) (-28.22%) (-27.76%) (-27.56%) (-26.66%) (-19.91%)

360th day Pm > Hc > Spc > CC > Cb > OL
(-26.58%) (-19.57%) (-18.82%) (-12.82%) (-8.96%) (-8.50%)

Ca^{+2} – ATPase activity (Tables. 11.1 & 11.2 Figs. 45 to 50)

Along with Na^{+}, K^{+} and Mg^{+2} – ATPase, Ca^{+2} – ATPase activity was also estimated in different regions of mice following prolonged treatment with galantamine hydrobromide. As in the case of Na^{+}, K^{+} and Mg^{+2} – ATPase, between 30th to 180th day On 30th day Ca^{+2} – ATPase activity also got inhibited in different regions of mice brain as follows.

30th day Spc > CC > Pm > Cb > Hc > OL
(-25.71%) (-20.77%) (-20.35%) (-18.79%) (-17.38%) (-16.40%)

60th day Pm > Spc > CC > Cb > OL > Hc
(-31.55%) (-30.81%) (-27.67%) (-26.28%) (-26.15%) (-27.67%)

90th day Pm > Spc > Cb > CC > OL > Hc
(-35.42%) (-35.40%) (-33.82%) (-32.93%) (-32.42%) (-32.04%)

120th day OL > Spc > Pm > CC > Cb > Hc
(-43.80%) (-43.75%) (-41.51%) (-42.53%) (-40.26%) (-33.57%)

150th Cb > OL > Spc > CC > Pm > Hc
(-50.03%) (-48.91%) (-47.46%) (-46.91%) (-45.45%) (-41.84%)

The **maximum inhibition** was noticed in all regions on 180th day except in Cerebellum.

180th day Spc > Pm > CC > Cb > OL > Hc
(-54.30%) (-53.45%) (-53.21%) (-51.99%) (-51.53%) (-48.81%)

From 210th day onwards up to 360th day Ca^{+2} – ATPases showed a gradual increasing trend.

210th day Spc > CC > Pm > OL > Cb > Hc
(-52.41%) (-51.79%) (-49.68%) (-47.21%) (-46.21%) (-41.78%)

240th day Pm > CC > Spc > OL > Cb > Hc
(-51.56%) (-45.27%) (-43.67%) (-41.88%) (-40.69%) (-39.92%)

270th day Pm > Spc > CC > Cb > OL > Hc
(-43.92%) (-39.84%) (-39.63%) (-36.83%) (-36.05%) (-35.04%)

300th day Pm > CC > Spc > OL > Cb > Hc
(-39.30%) (-35.76%) (-34.61%) (-31.70%) (-30.10%) (-28.49%)

330th day Pm > Spc > Hc > CC > Cb > OL
(-33.91%) (-28.03%) (-24.40%) (-21.42%) (-20.73%) (-16.42%)

360th day Pm > Hc > Spc > CC > OL > Cb
(-19.51%) (-16.25%) (-15.71%) (-12.26%) (-11.77%) (-9.40%)

An interesting observation was that the levels of all Na^+/K^+, Mg^{+2} and Ca^{+2} ATPases on 360th day and 15th day was similar indicating that the prolonged effect of galantamine hydrobromide getting reduced.

Discussion

In the present chapter the changes in the enzymes involved in energy metabolism and membrane transport functions, viz. Na^+/K^+, Mg^{+2} and Ca^{+2} ATPase activities were recorded in some selected regions of brain of mice at different time intervals following the oral administration of galantamine. One general observation in the present study was that all ATPases recorded significant inhibition in the brain areas of mice treated with galantamine hydrobromide for some period and after that all the ATPase levels showed gradual recovery.

The Na^+, K^+- ATPase or Na^+ pump, is an energy transducting ion pump first described by Skou in 1952 **(Skou, 1998)**. In recent years, research on Na^+, K^+- ATPase revealed that interactions of Na^+, K^+- ATPase with other proteins not only are important for regulation of pumping function, but also make it possible for the enzyme to function as a single transductor **(Xie and Cai, 2003)**. Long-term pharmacological interruption of cholinergic transmission can decrease the post synaptic membrane potential by altering Na^+, K^+- ATPase activity. This can be seen as a decline in [3H] ouabaine binding **(Henning *et al.*, 1994)**.

It has been reported that the enzymes like ATPases are very important for brain functions and diseases **(Jentsch *et al.*, 2000)**. There were also evidences that the cholinesterase inhibitors like galantamine hydrobromide, rivastigmine and seleginline act as inhibitors of Monoamine Oxidse, Acetylcholinesterase, Na^+, K^+- , Mg^{+2}- ATPase activities **(Haris Carageorgiou *et al.*, 2008)**.

Age - associated impairments in a test of attention and evidence of involvement of cholinergic system was referred by **Jones and Collegues (1995)**. It is known that inhibition of Na^+, K^+- ATPase induces neurotransmitter release in several experimental models **(Rodriguez de Lores Anaiz and Peltegrino de Iraldi, 1991)**. Further, studies revealed that Na^+, K^+- ATPase might play a role on memory formation **(Dos Reis-Lunardelli *et al.*, 2007)**. According to **Gorini and Collegues (2002)**

Observations in the present study were well supported by **Gorini and Collegues, 2002.** By the treatment of memory enhancing or Alzheimer's disease treated drugs decreased Mg^{+2}- ATPase activity in the frontal cortex of old aged rats was reported by **Gorini and Collegues (2002)**. In the present study also the decreased levels of Mg^{+2}- ATPase activity was observed upto 360 days in without Alzheimer's disease mice also.

During Alzheimer's disease the calcium release was enhanced and transported through ER calcium channels and the drugs which are used to treat AD were act as inhibitors of Endoplasmic Reticulum calcium channels and protects cell from apoptosis agents including Aβ **(Cedazo – Minguez *et al.*, 2002; Chan *et al.*, 2000).**

Calcium ions play several roles in processes which are crucial for the functioning of a neuron including membrane excitability and gene expression. Disturbances in calcium homeostasis in cells from Alzheimer's Disease patients have been observed for many years. Proper calcium signalling is crucial for synaptogenesis and dendritic spines plasticity. Cognitive decline and neurodegeneration of synapses are the first symptoms of Alzheimer's Disease **(Oertner and Matus, 2005; Redmond and Ghosh, 2005 and Hsieh *et al.*, 2006).** One of the recent study has proved that enhanced calcium release from Endoplasmic Reticulum calcium stores, altered synaptic transmission in hippocampal neurons that causes Alzheimer's disease **(Priller *et al.*, 2007).** Indeed the drugs that treat Alzheimer's Disease like galantamine hydrobromide, Memantine act as non-competitive antagonist of NMDA calcium channel receptors predominantly localized at dendritic spines belong to the group of very few drugs that have been proven to have some benefit to Alzheimer's patients **(Tariot *et al.*, 2004 and Peskind *et al.*, 2006).** These earlier reports further substantiate the results in the present study where inhibition in Ca^{+2}- ATPase activity levels under disease free conditions was observed on treatment with galantamine hydrobromide.

It was also reported that the galantamine hydrobromide has a unique character to inhibits the Na^{+}, K^{+}- Pump due to non specific reduction of Na^{+} influx which stimulates the Na^{+}-Ca^{+2} exchanger and inhibits K^{+} conductance **(Radicheva *et al.*, 1999).** There were also reports that in certain embodiments, the condition is hypoxia-induced dementia, such as Alzheimer's disease, anther aspect of the invention provides a pharmaceutical formulation comprising a Na^{+}/K^{+}- ATPase inhibitor, either alone or in combination with an agent effective for treating or preventing Alzheimer's disease, formulated in a pharmaceutically acceptable recipient and suitable for use in human patients to treat or prevent Alzheimer's disease **(Rodrigez and Pena, 1995).** Because of this the drugs used to treat for Alzheimer's Disease shows inhibitory effect on Na^{+}, K^{+}- ATPases in normal condition also.

Eaten and Salt (1989) have stated that in neurons from ventrobasal thalamus 5-HT enhances both NMDA and non-NMDA – mediated effects; such action of 5-HT, however is indirectly mediated by an inhibition of Na^{+}/K^{+} current **(Eaten and Salt, 1989)**. The cytoplasmic free Ca^{+2} is elevated in aged neurons and neurodegeneration encountered during Alzheimer's Disease **(Buj *et al.*, 2003; Geula *et al.*, 2003 and Nishimura *et al.*, 2005)**. So, the drugs used to treat Alzheimer's disease were shown to have inhibitory activity on Ca^{+2}- ATPase activity.

There was also evidence that the drugs act as Monoamine oxidase inhibitors also exhibit inhibitory effects on ATPases **(Altura, 1985; Samantaray *et al.*, 2003; Kosenko, 2003)**. It is the best support to the present study, where galantamine hydrobromide one of the MAO inhibitor also lowered ATPases activities in all regions of mice brain. So the drugs used to treat memory impairments have the character to decrease Ca^{+2}- ATPase levels and MAO levels also **(Heidrich *et al.*, 1997; Yamada *et al.*, 2000)**.

The results presented in this chapter indicating that the administration of galantamine hydrobromide decreases the activities of the ATPase enzymes related to energy metabolism and membrane transport functions in different regions and to different extent. The reason for different regions of the brain exhibiting different levels of sensitivity to galantamine hydrobromide treatment perhaps may be because of the heterogeneity of the brain and also the functions with which they are associated. For example olfactory lobe is implicated in the manifestation of sensory functions, hippocampus in memory and cognitive function, cerebral cortex in reflex actions and locomotor activities, cerebellum in locomotory functions and co-ordination of movements, pons medulla in respiratory functions and spinal cord in voluntary movements and co-ordination of muscles.

Table 9.1 : Changes in Na^+, K^+ - ATPase activity (µ moles of Pi formed/mg protein/h) in Olfactory Lobe (OL), Cerebellum (CB), Cerebral Cortex (CC) Hippocampus(Hc), Pons medulla (Pm) and Spinal Cord (SPC) of male albino mice on prolonged exposure to Galantamine Hydrobromide. Values in parenthesis indicate percent changes from control.

	CHRONIC											
	C	15 DAYS	C	30 DAYS	C	45 DAYS	C	60 DAYS	C	75 DAYS	C	90 DAYS
OL	26.38 ± 0.51	21.45** ± 0.35 (-18.60)	25.18 ±0.48	20.39** ±0.33 (-19.02)	27.18 ±0.35	18.56** ±0.30 (-31.71)	27.35 ±0.37	18.34** ±0.46 (-32.94)	27.38 ±0.54	17.31** ±0.28 (-36.77)	27.04 ±0.35	16.35** ±0.60 (-39.53)
CB	29.12 ±0.31	23.65** ±0.47 (-18.78)	28.16 ±0.34	22.12** ±0.46 (-21.44)	29.86 ±0.55	22.31** ±0.45 (-25.28)	27.75 ±0.52	19.27** ±0.44 (-30.55)	29.42 ±0.41	20.31** ±0.38 (-31.64)	29.11 ±0.46	19.25** ±0.60 (-33.87)
HC	31.81 ±0.53	26.09** ±0.44 (-17.98)	28.65 ±0.26	23.12** ±0.32 (-19.30)	28.31 ±0.37	22.41** ±0.34 (-20.84)	29.66 ±0.48	21.44** ±0.64 (-27.71)	27.44 ±0.44	18.26** ±0.21 (-33.45)	29.44 ±0.52	19.35** ±0.36 (-34.27)
CC	40.75 ±0.33	34.69** ±0.29 (-14.87)	41.33 ±0.32	32.10** ±0.57 (-22.33)	40.77 ±0.29	29.36** ±0.34 (-27.98)	36.12 ±0.43	25.26** ±0.56 (-30.06)	42.63 ±0.31	28.32** ±0.24 (-33.56)	37.05 ±0.36	23.45** ±0.26 (-36.70)
PM	37.12 ±0.52	29.66** ±0.28 (-20.09)	33.42 ±0.57	26.31** ±0.22 (-21.27)	34.91 ±0.55	23.85** ±0.49 (-31.68)	33.99 ±0.40	22.05** ±0.33 (-35.12)	31.18 ±0.35	19.35** ±0.28 (-37.94)	36.23 ±0.24	22.31** ±0.41 (38.42)
SPC	29.81 ±0.28	22.65** ±0.49 (-24.01)	30.02 ±0.41	21.52** ±0.37 (-28.31)	31.26 ±0.37	22.13** ±0.48 (-29.20)	29.66 ±0.40	19.25** ±0.42 (-35.09)	31.43 ±0.69	20.22** ±0.30 (-35.66)	31.11 ±0.58	19.65** ±0.43 (-36.83)

	C	105 DAYS	C	120 DAYS	C	135 DAYS	C	150 DAYS	C	165 DASYS	C	180 DAYS
OL	26.12 ±0.44	15.49** ±0.46 (-40.69)	27.32 ±0.31	14.65** ±0.22 (-46.37)	26.35 ±0.23	13.25** ±0.40 (-49.71)	26.41 ±0.69	12.44** ±0.46 (-52.89)	25.31 ±0.60	11.26** ±0.35 (-55.51)	26.36 ±0.72	11.25** ±0.33 (-57.32)
CB	27.44 ±0.45	17.05** ±0.67 (-37.86)	29.01 ±0.26	16.28** ±0.44 (-43.88)	29.07 ±0.42	14.21** ±0.31 (-51.11)	29.11 ±0.29	13.55** ±0.47 (-53.45)	27.64 ±0.39	12.14** ±0.40 (-56.87)	29.03 ±0.23	12.36** ±0.33 (-57.42)
HC	29.07 ±0.32	18.36** ±0.40 (-36.84)	29.85 ±0.25	18.35** ±0.33 (-38.52)	29.10 ±0.33	16.32** ±0.31 (-43.91)	31.05 ±0.26	16.27** ±0.55 (-47.76)	30.66 ±0.39	15.23** ±0.31 (-50.32)	28.55 ±0.20	12.69** ±0.32 (-55.55)
CC	41.88 ±0.46	25.39** ±0.62 (-39.37)	33.22 ±0.31	18.25** ±0.26 (-45.08)	36.44 ±0.53	19.25** ±0.38 (-47.17)	39.11 ±0.24	18.35** ±0.20 (-53.08)	42.31 ±0.33	17.26** ±0.36 (-59.20)	42.11 ±0.35	18.63** ±0.26 (-55.75)
PM	33.41 ±0.53	19.67** ±0.33 (-41.12)	35.13 ±0.46	18.96** ±0.44 (-46.02)	33.25 ±0.40	17.13** ±0.22 (-48.48)	35.11 ±0.38	17.36** ±0.56 (-50.55)	34.11 ±0.43	15.36** ±0.48 (-57.46)	32.18 ±0.26	11.66** ±0.49 (-63.76)
SPC	30.32 ±0.28	18.32** ±0.28 (-39.57)	30.16 ±0.44	15.31** ±0.21 (-49.23)	29.46 ±0.33	14.29** ±0.19 (-51.49)	29.73 ±0.31	14.25** ±0.18 (-52.06)	34.45 ±0.38	12.41** ±0.28 (-59.24)	29.91 ±0.40	12.62** ±0.31 (-57.80)

Values are mean ± SEM of six observations each from tissues pooled from 6 animals

**Values are significant at $p<0.01$

* Indicate significance at $p<0.05$

Not significant

Table 9.2 : Changes in Na^+, K^+ - ATPase activity (µ moles of Pi formed/mg protein/h) in Olfactory Lobe (OL), Cerebellum (CB), Cerebral Cortex (CC) Hippocampus(Hc), Pons medulla (Pm) and Spinal Cord (SPC) of male albino mice on prolonged exposure to Galantamine Hydrobromide. Values in parantheses indicate percent changes from control.

	CHRONIC																							
	C	195 DAYS	C	210 DAYS	C	225 DAYS	C	240 DAYS	C	255 DAYS	C	270 DAYS	C	285 DAYS	C	300 DAYS	C	315 DAYS	C	330 DAYS	C	345 DASYS	C	360 DAYS
OL	27.08 ± 0.28	12.25** ± 0.33 (-54.76)	26.43 ±0.19	12.35** ±0.53 (-53.27)	26.21 ±0.48	12.39** ±0.24 (-52.72)	25.18 ±0.38	13.56** ±0.33 (-46.14)	26.32 ±0.57	14.35** ±0.21 (-45.47)	25.73 ±0.83	15.69** ±0.46 (-39.02)	26.34 ±0.28	16.52** ±0.43 (-37.28)	25.11 ±0.69	16.33** ±0.18 (-34.96)	26.21 ±0.38	19.65** ±0.26 (-25.02)	26.18 ±0.28	21.32** ±0.44 (-18.56)	26.12 ±0.47	21.35** ±0.51 (-18.26)	26.88 ±0.34	24.68** ±0.40 (-8.18)
CB	27.65 ±0.40	12.33** ±0.38 (-55.40)	29.43 ±0.28	13.68** ±0.17 (-53.51)	29.76 ±0.48	15.21** ±0.33 (-48.89)	27.85 ±0.40	15.22** ±0.34 (-45.35)	29.11 ±0.34	17.45** ±0.43 (-40.05)	30.41 ±0.62	17.22** ±0.39 (-43.37)	29.12 ±0.33	15.56** ±0.29 (-46.56)	29.03 ±0.41	19.27** ±0.58 (-33.62)	27.45 ±0.58	19.32** ±0.31 (-29.61)	29.11 ±0.32	21.45** ±0.44 (-26.31)	29.12 ±0.31	23.34** ±0.42 (-19.84)	29.21 ±0.19	25.62** ±0.42 (-12.29)
HC	31.22 ±0.37	14.36** ±0.35 (-54)	30.95 ±0.54	16.32** ±0.28 (-47.26)	30.01 ±0.50	16.22** ±0.25 (-45.95)	31.16 ±0.44	19.25** ±0.35 (-38.22)	28.03 ±0.33	15.41** ±0.48 (-44.98)	29.24 ±0.28	19.33** ±0.20 (-33.89)	30.43 ±0.27	19.32** ±0.25 (-36.51)	27.46 ±0.43	18.35** ±0.40 (-33.17)	27.14 ±0.28	19.25** ±0.24 (-29.07)	29.11 ±0.38	21.31** ±0.37 (-26.79)	27.45 ±0.22	21.35** ±0.76 (-22.22)	29.12 ±0.26	23.58** ±0.38 (-19.02)
CC	42.68 ±0.36	19.05** ±0.38 (-55.36)	41.26 ±0.55	19.35** ±0.31 (-53.10)	41.85 ±0.18	19.35** ±0.11 (-52.06)	41.13 ±0.35	18.36** ±0.17 (-55.36)	33.15 ±0.37	16.28** ±0.36 (-50.88)	38.33 ±0.21	21.36** ±0.26 (-44.27)	37.26 ±0.35	21.68** ±0.17 (-41.81)	29.24 ±0.36	17.36** ±0.25 (-40.62)	31.65 ±0.34	21.56** ±0.32 (-31.87)	39.25 ±0.18	29.05** ±0.41 (-25.98)	35.25 ±0.46	28.65** ±0.35 (-18.72)	41.39 ±0.71	36.12** ±0.30 (-12.73)
PM	34.11 ±0.43	13.68** ±0.33 (-59.89)	38.55 ±0.34	16.35** ±0.36 (-57.58)	33.26 ±0.28	15.38** ±0.21 (-53.75)	38.33 ±0.36	18.36** ±0.33 (-52.10)	34.66 ±0.55	16.25** ±0.19 (-53.11)	35.25 ±0.55	18.35** ±0.26 (-47.94)	35.32 ±0.41	18.65** ±0.34 (-47.19)	37.66 ±0.37	21.68** ±0.38 (-42.43)	33.25 ±1.05	20.35** ±0.72 (-38.79)	34.10 ±0.55	21.55** ±0.48 (-36.80)	35.06 ±0.30	25.65** ±0.61 (-26.83)	33.99 ±0.30	25.14** ±0.55 (-26.03)
SPC	30.25 ±0.42	12.96** ±0.59 (-57.15)	31.32 ±0.42	14.18** ±0.39 (-54.72)	29.84 ±0.37	14.27** ±0.37 (-52.17)	29.46 ±0.46	15.24** ±0.42 (-48.26)	30.15 ±0.29	16.25** ±0.45 (-46.10)	29.18 ±0.34	16.27** ±0.32 (-44.24)	31.46 ±0.32	19.34** ±0.27 (-38.52)	30.05 ±0.33	19.22** ±0.33 (-36.03)	30.75 ±0.51	21.32** ±0.22 (-30.66)	30.85 ±0.51	22.54** ±0.49 (-26.93)	29.21 ±0.41	21.39** ±0.53 (-26.77)	30.02 ±0.59	24.68** ±0.29 (-17.78)

Values are mean ± SEM of six observations each from tissues pooled from 6 animals

**Values are significant at $p<0.01$

* Indicate significance at $p<0.05$

\# Not significant

Table 10.1 : Changes in Mg^{+2}- ATPase activity (μ moles of Pi formed/mg protein/h) in Olfactory Lobe (OL), Cerebellum (CB), Cerebral Cortex (CC) Hippocampus(Hc), Pons medulla (Pm) and Spinal Cord (SPC) of male albino mice on prolonged exposure to Galantamine Hydrobromide. Values in parenthesis indicate percent changes from control.

CHRONIC												
	C	15 DAYS	C	30 DAYS	C	45 DAYS	C	60 DAYS	C	75 DAYS	C	90 DAYS
OL	29.06 ± 0.43	24.32** ± 0.44 (-16.31)	28.16 ±0.37	23.54** ±0.67 (-16.40)	29.15 ±0.40	22.25** ±0.53 (-23.67)	29.32 ±0.73	21.65** ±0.28 (-26.15)	29.41 ±0.45	20.35** ±0.54 (-30.80)	29.08 ±0.36	19.65** ±0.51 (-32.42)
CB	31.08 ±0.60	25.31** ±0.66 (-18.56)	31.18 ±0.43	25.32** ±0.60 (-18.79)	31.83 ±0.47	25.36** ±0.43 (-20.37)	31.54 ±0.44	23.25** ±0.50 (-26.28)	30.76 ±0.47	21.58** ±0.36 (-29.84)	32.11 ±0.54	21.25** ±0.39 (-33.82)
HC	34.18 ±0.68	29.32** ±0.44 (-11.29)	31.86 ±0.33	26.32** ±0.33 (-17.38)	31.66 ±0.50	25.47** ±0.55 (-15.55)	32.69 ±0.55	24.59** ±0.48 (-24.77)	31.48 ±0.50	21.59** ±0.41 (-31.41)	31.86 ±0.60	21.65** ±0.47 (-32.04)
CC	43.78 ±0.38	37.12** ±0.47 (-15.21)	44.38 ±0.25	35.16** ±0.48 (-20.77)	42.76 ±0.34	32.65** ±0.44 (-23.64)	39.21 ±0.30	28.36** ±0.36 (-27.67)	45.65 ±0.58	31.12** ±0.43 (-31.82)	39.11 ±0.39	26.23** ±0.32 (-32.93)
PM	40.22 ±0.62	32.68** ±0.56 (-18.74)	36.85 ±0.33	29.35** ±0.33 (-28.35)	37.98 ±0.53	28.45** ±0.28 (-25.09)	36.98 ±0.44	25.31** ±0.80 (-31.55)	33.29 ±0.40	22.56** ±0.51 (-32.23)	39.32 ±0.54	25.39** ±0.49 (-35.42)
SPC	32.86 ±0.62	25.61** ±0.57 (-22.86)	33.09 ±0.48	24.58** ±0.27 (-25.71)	34.62 ±0.46	25.34** ±0.72 (-26.80)	32.65 ±0.37	22.59** ±0.39 (-30.81)	34.56 ±0.32	23.58** ±0.39 (-31.77)	33.41 ±0.44	21.58** ±0.48 (-35.40)

CHRONIC												
	C	105 DAYS	C	120 DAYS	C	135 DAYS	C	150 DAYS	C	165 DASYS	C	180 DAYS
OL	28.33 ±0.38	18.63** ±0.40 (-34.23)	29.15 ±0.65	16.38** ±0.30 (-43.80)	28.63 ±0.39	15.68** ±0.36 (-45.27)	28.11 ±0.54	14.36** ±0.33 (-48.91)	28.75 ±0.40	14.32** ±0.45 (-50.19)	29.55 ±0.53	14.32** ±0.25 (-51.53)
CB	32.05 ±0.46	19.65** ±0.32 (-38.68)	30.67 ±0.55	18.32** ±0.43 (-40.26)	32.05 ±0.33	17.35** ±0.25 (-45.86)	30.66 ±0.43	15.32** ±0.49 (-50.03)	30.68 ±0.39	15.23** ±0.40 (-50.35)	32.06 ±0.34	15.39** ±0.24 (-51.99)
HC	30.88 ±0.50	21.36** ±0.48 (-30.82)	32.14 ±0.47	21.35** ±0.25 (-33.57)	32.15 ±0.45	19.65** ±0.57 (-38.88)	33.18 ±0.53	19.29** ±0.46 (-41.84)	33.65 ±0.39	18.36** ±0.35 (-45.43)	34.48 ±0.36	17.65** ±0.37 (-48.81)
CC	44.89 ±0.46	29.56** ±0.28 (-34.15)	38.65 ±0.51	22.21** ±0.54 (-42.53)	36.32 ±0.40	20.05** ±0.55 (-44.79)	42.18 ±0.25	22.39** ±0.28 (-46.91)	45.78 ±0.37	22.14** ±0.89 (-51.63)	45.66 ±0.43	21.36** ±0.50 (-53.21)
PM	36.45 ±0.32	22.39** ±0.33 (-38.57)	36.52 ±0.43	21.36** ±0.31 (-41.51)	38.41 ±0.61	21.59** ±0.38 (-43.79)	41.38 ±0.46	22.57** ±0.46 (-45.45)	35.29 ±0.51	14.65** ±0.45 (-58.48)	41.51 ±0.53	19.32** ±0.44 (-53.45)
SPC	34.18 ±0.40	21.28** ±0.40 (-37.74)	33.12 ±0.37	18.63** ±0.28 (-43.75)	31.76 ±0.54	17.65** ±0.40 (-44.42)	32.76 ±0.41	17.21** ±0.43 (-47.46)	31.96 ±0.28	15.68** ±0.31 (-50.93)	33.68 ±0.53	15.39** ±0.33 (-54.30)

Values are mean ± SEM of six observations each from tissues pooled from 6 animals

**Values are significant at $p<0.01$

* Indicate significance at $p<0.05$

\# Not significant

Table 10.2 : Changes in Mg^{+2}- ATPase activity (μ moles of Pi formed/mg protein/h) in Olfactory Lobe (OL), Cerebellum (CB), Cerebral Cortex (CC) Hippocampus(Hc), Pons medulla (Pm) and Spinal Cord (SPC) of male albino mice on prolonged exposure to Galantamine Hydrobromide. Values in parenthesis indicate percent changes from control.

CHRONIC																								
	C	195 DAYS	C	210 DAYS	C	125 DAYS	C	240 DAYS	C	255 DAYS	C	270 DAYS	C	285 DAYS	C	300 DAYS	C	315 DAYS	C	330 DAYS	C	345 DASYS	C	360 DAYS
OL	29.45 ±0.25	15.31** ±0.64 (-48.01)	29.10 ±0.26	15.36** ±0.21 (-47.21)	28.44 ±0.26	15.37** ±0.23 (-45.95)	28.17 ±1.38	16.37** ±0.42 (-41.88)	29.58 ±0.31	17.68** ±0.31 (-40.22)	28.76 ±0.27	18.39** ±0.31 (-36.05)	28.66 ±0.29	19.37** ±0.33 (-32.41)	28.32 ±0.30	19.34** ±0.31 (-31.70)	28.32 ±0.24	22.41** ±0.31 (-20.86)	29.16 ±0.25	24.37** ±0.22 (-16.42)	29.11 ±0.45	24.39** ±0.17 (-16.21)	29.89 ±0.22	26.37** ±0.32 (-11.77)
CB	30.16 ±0.35	16.52** ±0.23 (-54.77)	31.56 ±0.31	16.97** ±0.66 (-46.21)	31.86 ±0.40	18.34** ±0.34 (-42.43)	34.96 ±0.37	18.36** ±0.37 (-40.69)	33.01 ±0.32	20.38** ±0.66 (-38.26)	31.11 ±0.31	19.65** ±0.34 (-36.83)	31.13 ±0.38	20.39** ±0.51 (-34.50)	30.56 ±0.53	21.36** ±0.47 (-30.10)	31.09 ±0.19	22.45** ±0.49 (-17.79)	31.15 ±0.54	24.69** ±0.5 (-20.73)	31.56 ±0.44	26.58** ±0.47 (-15.77)	31.25 ±0.48	28.31** ±0.21 (-9.40)
HC	33.96 ±0.29	19.32** ±0.44 (-43.10)	31.52 ±0.58	18.38** ±0.41 (-41.78)	33.11 ±0.46	19.65** ±0.37 (-40.65)	31.11 ±0.29	18.69** ±0.22 (-39.92)	33.85 ±0.31	21.36** ±0.44 (-36.89)	33.19 ±0.38	21.56** ±0.56 (-35.04)	32.58 ±0.40	22.69** ±0.51 (-30.47)	30.18 ±0.30	21.58** ±0.43 (-28.49)	30.86 ±0.33	22.58** ±0.30 (-26.83)	32.16 ±0.28	24.31** ±0.28 (-24.40)	30.98 ±0.51	24.57** ±0.71 (-20.69)	32.18 ±0.47	26.95** ±0.66 (-16.25)
CC	45.12 ±0.35	21.19** ±0.40 (-53.03)	44.31 ±0.42	21.36** ±0.60 (-51.79)	43.11 ±0.33	22.69** ±0.55 (-47.36)	43.29 ±0.30	23.69** ±0.33 (-45.27)	36.32 ±0.37	20.89** ±0.33 (-42.48)	41.38 ±0.32	24.98** ±0.73 (-39.63)	39.62 ±0.35	24.85** ±0.53 (-37.27)	32.15 ±0.37	20.65** ±0.25 (-35.76)	33.79 ±0.48	24.56** ±0.51 (-27.31)	41.59 ±0.50	32.68** ±0.28 (-21.42)	38.54 ±0.42	31.29** ±0.31 (-18.81)	44.69 ±0.35	39.21** ±0.67 (-12.26)
PM	39.15 ±0.37	18.65** ±0.40 (-52.34)	36.29 ±0.54	18.26** ±0.28 (-49.68)	37.61 ±0.47	19.65** ±0.22 (-47.75)	37.12 ±1.06	17.98** ±0.80 (-51.56)	38.12 ±0.49	20.56** ±0.44 (-46.06)	38.61 ±0.57	21.65** ±0.79 (-43.92)	38.51 ±0.33	22.36** ±0.56 (-41.93)	40.61 ±0.56	24.65** ±0.56 (-39.30)	37.62 ±0.59	23.65** ±0.44 (-37.13)	37.21 ±0.51	24.99** ±0.98 (-33.91)	38.11 ±0.35	28.36** ±0.40 (-25.58)	35.98 ±0.55	28.96** ±0.48 (-19.51)
SPC	34.39 ±0.53	16.23** ±0.43 (-52.80)	33.54 ±0.43	15.96** ±0.32 (-52.41)	32.85 ±0.34	17.58** ±0.41 (-46.48)	32.56 ±0.35	18.34** ±0.31 (-43.67)	33.11 ±0.51	19.36** ±0.55 (-41.52)	32.15 ±0.48	19.34** ±0.26 (-39.84)	34.65 ±0.44	21.65** ±0.66 (-37.51)	33.11 ±0.51	41.65** ±0.60 (-34.61)	33.08 ±0.67	23.58** ±0.34 (-28.71)	33.85 ±0.41	24.36** ±0.43 (-28.03)	31.44 ±0.36	24.36** ±0.32 (-22.51)	33.08 ±0.54	27.88** ±0.54 (-15.71)

Values are mean ± SEM of six observations each from tissues pooled from 6 animals

**Values are significant at $p < 0.01$

* Indicate significance at $p<0.05$

Not significant

Table 11.1 : Changes in Ca^{+2}- ATPase activity (µ moles of Pi formed/mg protein/h) in Olfactory Lobe (OL), Cerebellum (CB), Cerebral Cortex (CC) Hippocampus(Hc), Pons medulla (Pm) and Spinal Cord (SPC) of male albino mice on prolonged exposure to Galantamine Hydrobromide. Values in parenthesis indicate percent changes from control.

	CHRONIC																							
	C	15 DAYS	C	30 DAYS	C	45 DAYS	C	60 DAYS	C	75 DAYS	C	90 DAYS	C	105 DAYS	C	120 DAYS	C	135 DAYS	C	150 DAYS	C	165 DASYS	C	180 DAYS
OL	24.11 ±0.60	19.38** ±0.48 (-19.61)	25.31 ±0.51	20.29** ±0.36 (-19.83)	26.08 ±0.28	17.32** ±0.38 (-33.58)	26.26 ±0.39	17.29** ±0.44 (-34.31)	26.30 ±0.37	16.29** ±0.48 (-38.06)	26.02 ±0.47	15.26** ±0.31 (-41.35)	25.10 ±0.66	14.36** ±0.31 (-42.78)	26.44 ±0.65	13.26** ±0.42 (-49.84)	25.30 ±0.31	12.21** ±0.24 (-51.73)	25.33 ±0.31	11.26** ±0.36 (-55.54)	24.16 ±0.66	10.32** ±0.45 (-57.28)	25.61 ±0.47	10.18** ±0.24 (-60.24)
CB	28.08 ±0.51	22.56** ±0.31 (-19.68)	27.11 ±0.26	21.05** ±0.55 (-22.35)	28.75 ±0.37	21.31** ±0.28 (-25.87)	26.54 ±0.26	18.25** ±0.21 (-31.23)	28.31 ±0.24	19.22** ±0.33 (-32.10)	28.06 ±0.36	18.11** ±0.36 (-35.45)	26.35 ±0.29	16.00** ±0.35 (-39.27)	28.06 ±0.33	15.23** ±0.17 (-45.72)	28.06 ±0.39	13.11** ±0.13 (-53.27)	28.05 ±0.35	12.53** ±0.31 (-55.32)	26.61 ±0.26	11.22** ±0.36 (-57.83)	26.54 ±0.22	11.13** ±0.38 (-58.06)
HC	30.75 ±0.31	25.06** ±0.26 (-18.50)	27.65 ±0.58	22.06** ±0.37 (-20.21)	27.30 ±0.36	21.35** ±0.39 (-21.79)	28.65 ±0.23	20.35** ±0.45 (-28.97)	26.43 ±0.53	17.35** ±0.22 (-34.35)	28.21 ±0.38	18.29** ±0.31 (-35.16)	28.01 ±0.26	17.32** ±0.31 (-38.16)	28.73 ±0.38	17.25** ±0.45 (-39.95)	28.05 ±0.29	15.31** ±0.26 (-45.41)	30.02 ±0.48	15.12** ±0.33 (-49.53)	29.38 ±0.44	14.12** ±0.30 (-51.94)	27.26 ±0.20	11.35**, ±0.40 (-58.36)
CC	39.51 ±0.34	33.65** ±0.35 (-14.83)	40.13 ±0.31	31.05** ±0.60 (-22.63)	39.54 ±0.31	28.14** ±0.15 (-28.83)	35.06 ±0.40	24.21** ±0.45 (-30.94)	41.44 ±0.28	27.12** ±0.36 (-34.55)	36.03 ±0.50	22.39** ±0.34 (-37.85)	40.81 ±0.32	24.13** ±0.30 (-40.87)	32.20 ±0.24	17.11** ±0.37 (-46.86)	35.32 ±0.44	18.21** ±0.25 (-48.44)	38.06 ±0.25	17.42** ±0.52 (-54.23)	41.67 ±0.37	18.21** ±0.17 (-56.29)	41.30 ±0.66	16.12** ±0.37 (-60.96)
PM	36.06 ±0.40	28.36** ±0.31 (-21.35)	32.40 ±0.31	25.29** ±0.40 (-21.94)	33.85 ±0.38	22.45** ±0.22 (-33.67)	32.95 ±0.27	21.02** ±0.40 (-36.20)	30.15 ±0.36	18.27** ±0.22 (-39.40)	35.26 ±0.40	21.22** ±0.37 (-40.19)	32.21 ±0.33	18.56** ±0.22 (-42.37)	34.11 ±0.60	17.59** ±0.30 (-48.43)	32.27 ±0.35	16.10** ±0.29 (-50.10)	34.15 ±0.42	16.25** ±0.39 (-52.41)	37.42 ±0.18	17.29** ±0.63 (-53.79)	31.17 ±0.91	10.68** ±0.28 (-65.73)
SPC	28.57 ±0.33	21.45** ±0.22 (-24.92)	29.05 ±0.47	20.55** ±0.48 (-29.25)	30.21 ±0.47	21.16** ±0.34 (-29.95)	28.54 ±0.36	18.23** ±0.24 (-36.12)	29.12 ±0.59	17.54** ±0.33 (-39.76)	30.38 ±0.55	19.21** ±0.21 (-36.76)	30.16 ±0.43	18.56** ±0.40 (-38.46)	29.11 ±0.24	14.39** ±0.29 (-50.56)	28.36 ±0.33	13.21** ±0.39 (-53.42)	28.63 ±0.30	13.22** ±0.34 (-53.82)	28.71 ±0.30	11.65** ±0.27 (-59.42)	29.42 ±0.26	11.29** ±0.25 (-61.62)

Values are mean ± SEM of six observations each from tissues pooled from 6 animals

**Values are significant at $p<0.01$

* Indicate significance at $p<0.05$

Not significant

Table 11.2 : Changes in Ca^{+2}- ATPase activity (µ moles of Pi formed/mg protein/h) in Olfactory Lobe (OL), Cerebellum (CB), Cerebral Cortex (CC) Hippocampus(Hc), Pons medulla (Pm) and Spinal Cord (SPC) of male albino mice on prolonged exposure to Galantamine Hydrobromide. Values in parenthesis indicate percent changes from control.

	CHRONIC																							
	C	195 DAYS	C	210 DAYS	C	225 DAYS	C	240 DAYS	C	255 DAYS	C	270 DAYS	C	285 DAYS	C	300 DAYS	C	315 DAYS	C	330 DAYS	C	345 DASYS	C	360 DAYS
OL	26.03 ± 0.26	11.15** ± 0.39 (-57.16)	25.41 ±0.33	11.26** ±0.40 (-55.68)	25.12 ±0.80	11.32** ±0.57 (-54.93)	24.16 ±0.25	12.56** ±0.31 (-48.01)	25.11 ±0.56	13.29** ±0.10 (-47.03)	24.68 ±0.35	19.39** ±0.28 (-41.69)	25.33 ±0.56	15.41** ±0.40 (-39.16)	24.05 ±0.32	15.26** ±0.36 (-36.54)	25.16 ±0.19	18.23** ±0.29 (-27.54)	25.16 ±0.49	20.15** ±0.42 (-19.91)	25.16 ±0.25	20.39** ±0.35 (-18.95)	25.86 ±0.42	23.16** ±0.43 (-8.50)
CB	28.01 ±0.26	11.29** ±0.18 (-59.68)	28.32 ±0.40	12.50** ±0.28 (-55.86)	28.72 ±0.48	14.19** ±0.40 (-50.59)	28.10 ±0.17	14.55** ±0.35 (-48.22)	26.65 ±0.27	14.15** ±0.30 (-46.90)	29.23 ±0.37	16.12** ±0.47 (-44.85)	28.05 ±0.26	16.44** ±0.41 (-41.39)	28.04 ±0.30	18.25** ±0.36 (-34.91)	26.25 ±0.15	18.26** ±0.32 (-30.43)	28.15 ±0.35	20.39** ±0.34 (-27.56)	28.16 ±0.31	22.33** ±0.35 (-20.70)	27.12 ±0.28	24.69** ±0.53 (-8.96)
HC	30.11 ±0.39	13.29** ±0.21 (-55.86)	29.95 ±0.48	15.27** ±0.45 (-49.01)	29.06 ±0.36	15.16** ±0.38 (-47.83)	27.02 ±0.35	14.29** ±0.34 (-47.11)	30.12 ±0.36	18.16** ±0.22 (-39.70)	29.23 ±0.34	18.32** ±0.28 (-37.32)	28.22 ±0.21	18.29** ±0.26 (-35.18)	26.42 ±0.52	17.29** ±0.25 (-34.55)	26.16 ±0.16	18.22** ±0.22 (-30.35)	28.09 ±0.29	20.29** ±0.31 (-27.76)	27.43 ±0.24	20.18** ±0.53 (-26.43)	28.09 ±0.27	22.59** ±0.38 (-19.57)
CC	41.00 ±0.42	17.26** ±0.60 (-57.90)	40.11 ±0.57	17.26** ±0.42 (-56.96)	32.11 ±0.50	15.21** ±0.16 (-52.63)	40.06 ±0.36	17.26** ±0.30 (-56.91)	40.21 ±0.29	18.32** ±0.25 (-54.43)	37.30 ±1.13	20.15** ±0.32 (-45.97)	36.22 ±0.34	20.65** ±0.38 (-42.98)	27.21 ±0.57	16.33** ±0.38 (-39.98)	30.55 ±0.49	20.45** ±0.53 (-33.06)	38.21 ±0.26	28.02** ±0.44 (-26.66)	34.31 ±0.60	27.56** ±0.43 (-19.43)	40.31 ±0.40	35.14** ±0.33 (-12.82)
PM	35.07 ±0.32	14.33** ±0.31 (-59.13)	37.45 ±0.50	15.33** ±0.40 (-59.10)	32.33 ±0.38	14.32** ±0.48 (-55.56)	33.15 ±0.37	12.65** ±0.28 (-61.84)	34.21 ±0.36	17.26** ±0.42 (-49.54)	34.21 ±0.41	17.56** ±0.29 (-48.66)	36.62 ±0.25	20.58** ±0.59 (-43.80)	32.52 ±0.31	18.95** ±0.30 (-41.72)	33.11 ±0.33	20.54** ±0.39 (-37.96)	33.52 ±0.40	23.52** ±0.42 (-29.80)	34.08 ±0.53	24.55** ±0.27 (-27.96)	32.95 ±0.35	24.19** ±0.56 (-26.58)
SPC	29.22 ±0.68	11.96** ±0.27 (-59.06)	30.23 ±0.40	13.14** ±0.10 (-56.53)	28.67 ±0.25	13.22** ±0.37 (-53.00)	28.42 ±0.31	14.12** ±0.40 (-50.31)	29.11 ±0.42	15.12** ±0.39 (-48.05)	28.19 ±0.42	15.21** ±0.25 (-46.04)	30.41 ±0.48	18.22** ±0.13 (-40.08)	29.03 ±0.34	18.11** ±0.12 (-37.61)	29.54 ±0.31	20.31** ±0.36 (-31.24)	29.76 ±0.24	21.36** ±0.28 (-28.22)	28.16 ±0.25	20.29** ±0.37 (-27.94)	29.05 ±0.22	23.58** ±0.33 (-18.82)

Values are mean ± SEM of six observations each from tissues pooled from 6 animals

**Values are significant at $p<0.01$

* Indicate significance at $p<0.05$

Not significant

Figs. 33,34&35 : Graphical representation of percent changes in the content of Na^{+}, K^{+}-ATPase (In vivo) in Olfactory Lobe (OL), Hippocampus (Hc), Cerebral Cortex (CC) of experimental mice brain following prolonged exposure to galantamine hydrobromide against the control.

Fig. 33

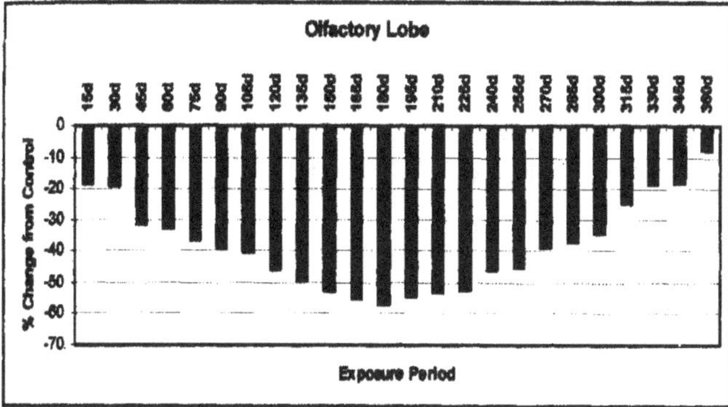

Fig. 34

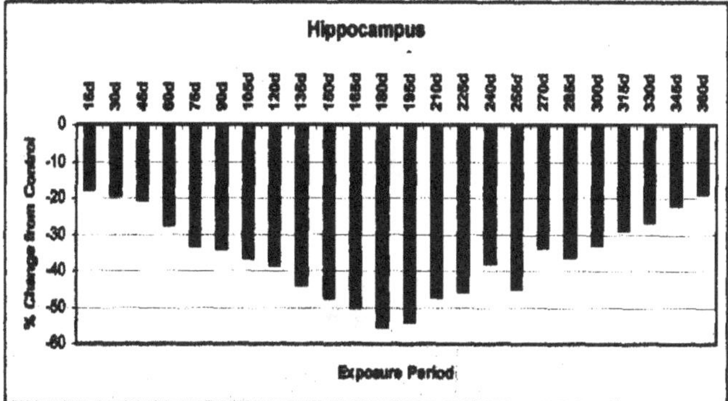

Fig. 35

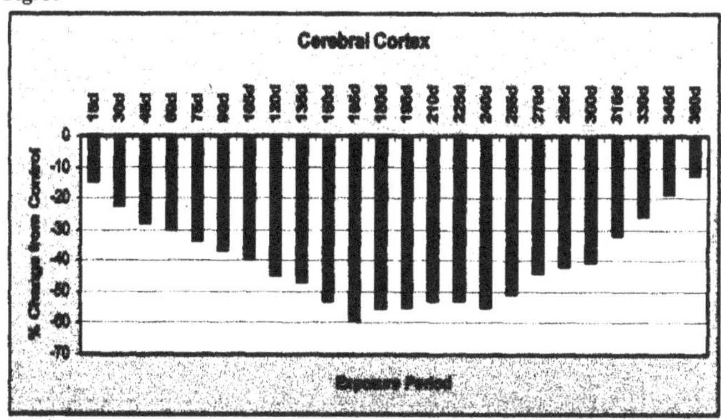

Figs. 36,37&38 : Graphical representation of percent changes in the content of Na^+, K^+ -ATPase (In vivo) in Cerebellum (Cb), Ponsmedulla (Pm), Spinal cord (Spc) of experimental mice brain following prolonged exposure to galantamine hydrobromide against the control.

Fig. 36

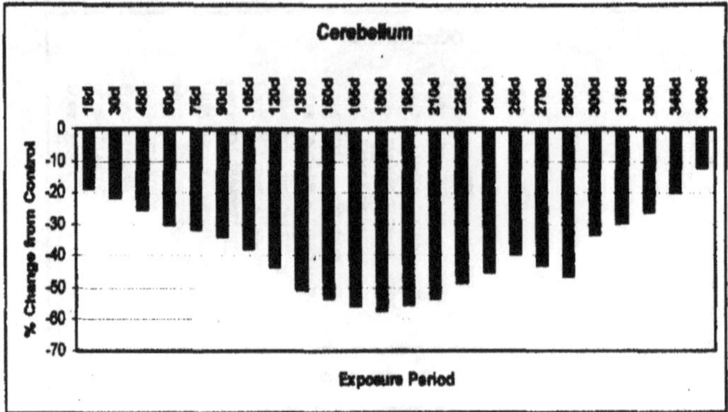

Fig. 37

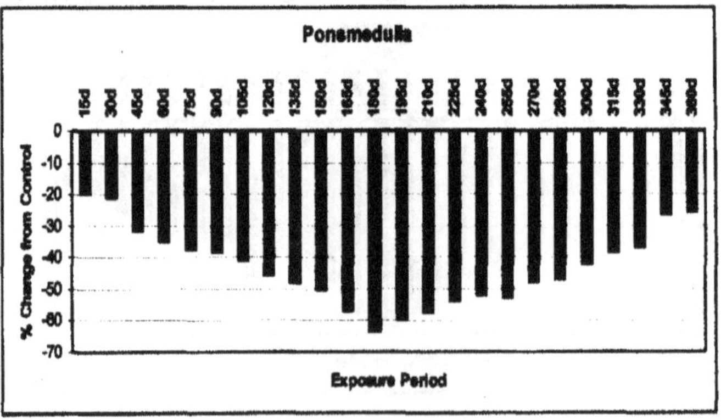

Fig. 38

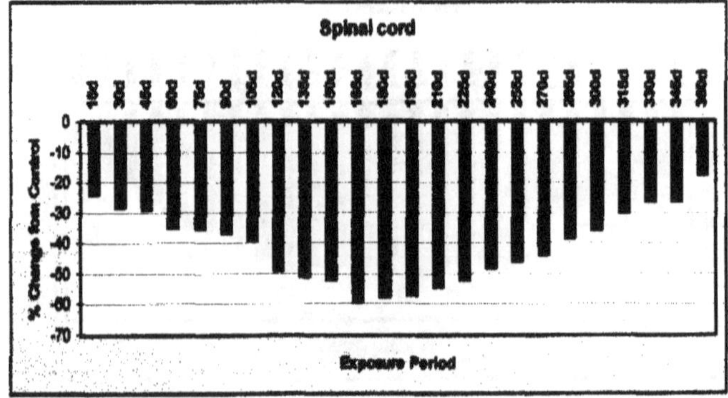

Figs. 39,40&41 : Graphical representation of percent changes in the content of Mg^{+2}-ATPase (In vivo) in Olfactory Lobe (OL), Hippocampus (Hc), Cerebral Cortex (CC) of experimental mice brain following prolonged exposure to galantamine hydrobromide against the control.

Fig. 39

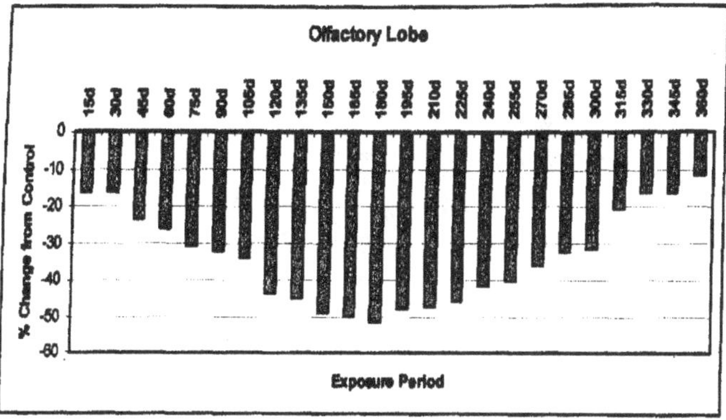

Fig. 40

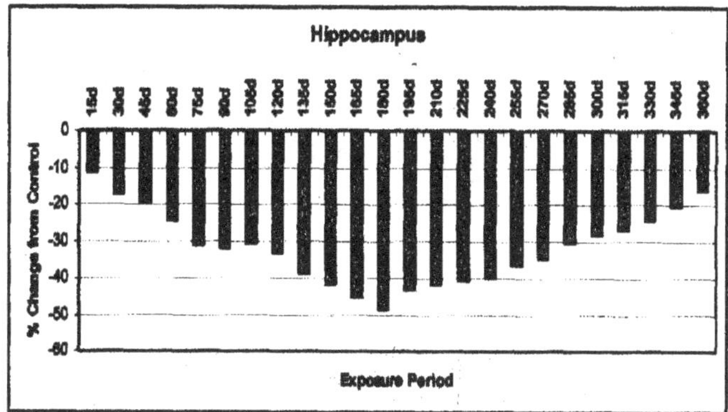

Fig. 41

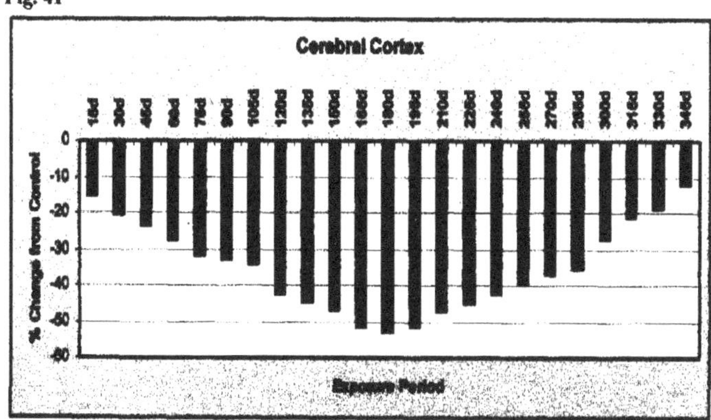

Figs. 42,43&44 : Graphical representation of percent changes in the content of Mg^{+2}-ATPase (In vivo) in Cerebellum (Cb), Ponsmedulla (Pm), Spinal cord (Spc) of experimental mice brain following prolonged exposure to galantamine hydrobromide against the control.

Fig. 42

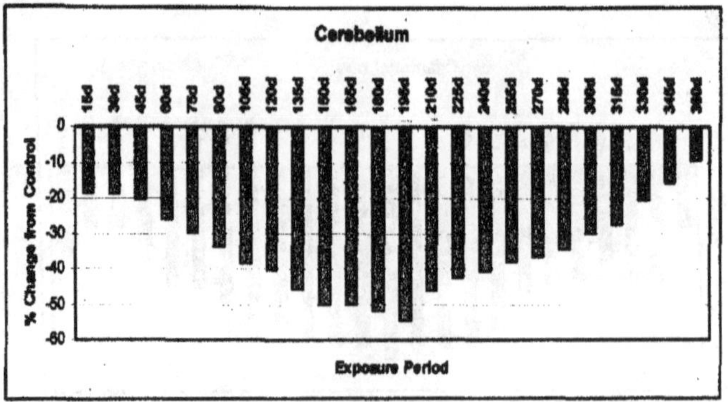

Fig. 43

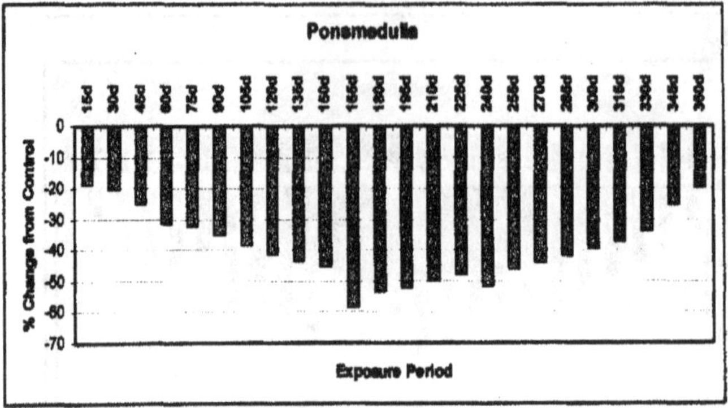

Fig. 44

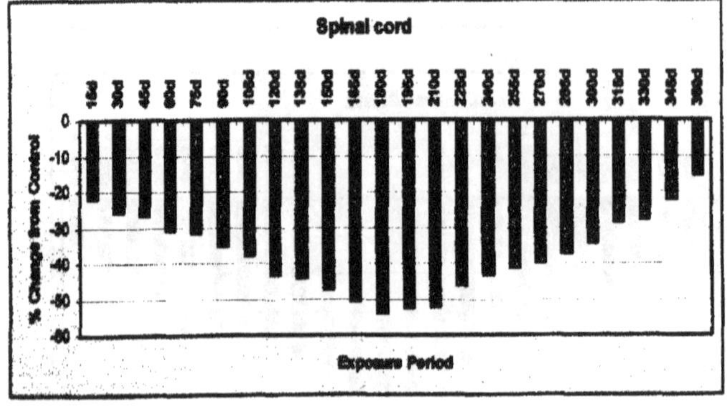

Figs. 45,46&47 : Graphical representation of percent changes in the content of Ca^{+2} - ATPase (In vivo) in Olfactory Lobe (OL), Hippocampus (Hc), Cerebral Cortex (CC) of experimental mice brain following prolonged exposure to galantamine hydrobromide against the control.

Fig. 45

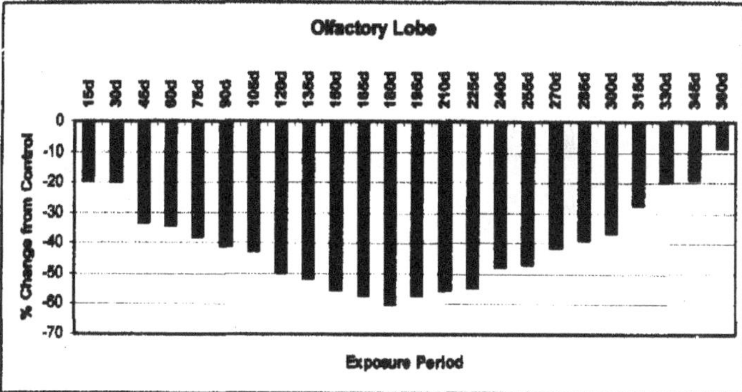

Fig. 46

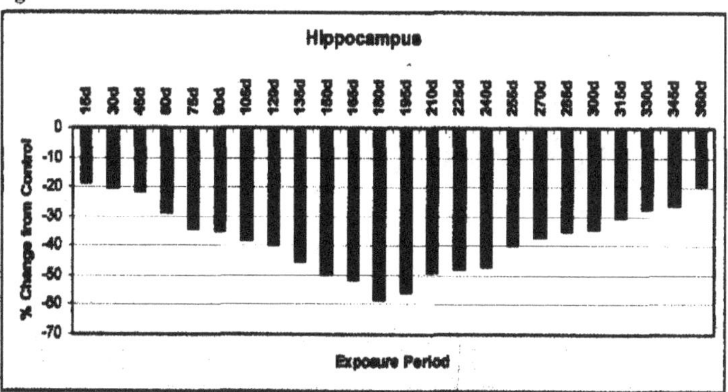

Fig. 47

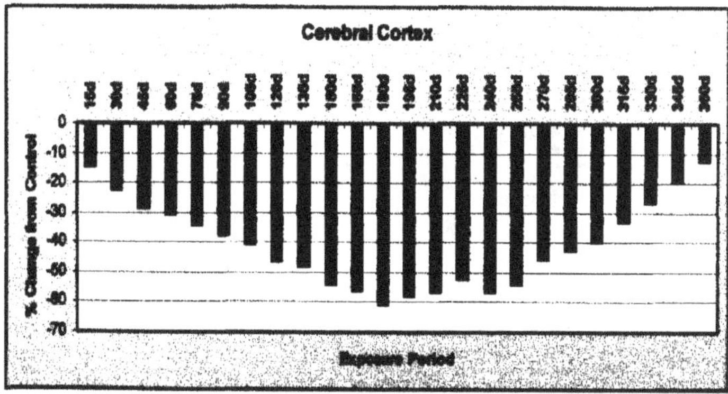

Figs. 48,49&50 : Graphical representation of percent changes in the content of Ca^{+2} - ATPase (In vivo) in Cerebellum (Cb), Ponsmedulla (Pm), Spinal cord (Spc) of experimental mice brain following prolonged exposure to galantamine hydrobromide against the control.

Fig. 48

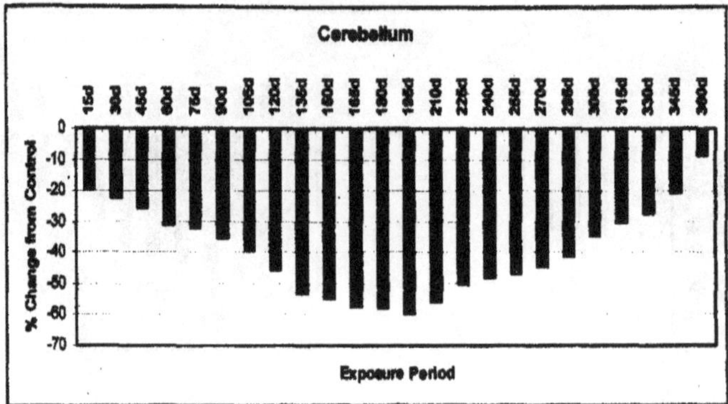

Fig. 49

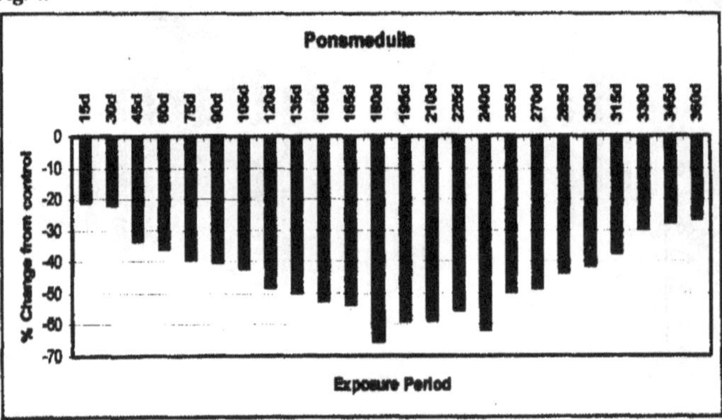

Fig. 50

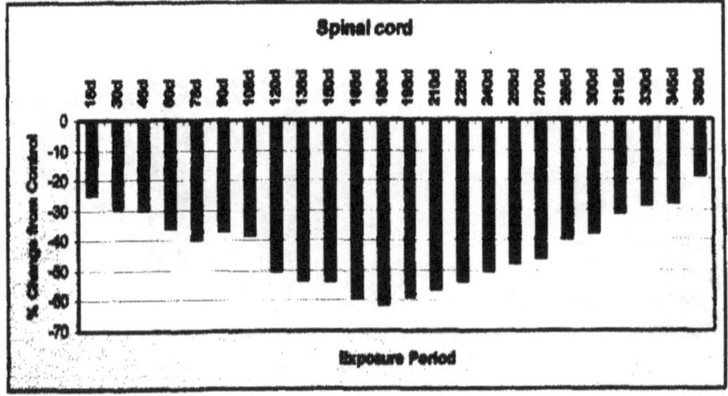

CHAPTER - IV

APPLICATION OF LIPINSKI RULE OF FIVE

In 1997 Christopher A. Lipinski published a seminal paper identifying a series of features commonly found in orally active drugs. These features are referred to as Lipinski's rule of five or a rule of thumb to evaluate druglikeness or determine a chemical compound with a certain pharmacological or biological activity that would make it a likely orally active drug in humans. The rule was formulated by A. Lipinski, based on the observation that most medication drugs are relatively small and lipophilic molecules **(Lipinski, 1997 and 2001).**

Dr. Christopher Lipinski, member of the T.B. Alliance Scientific Advisory Committee and a retired Senior Research Fellow at Pfizer Global R&D Groton, New London Labs, has been honoured for his ground breaking "Rule of Five" by the American Chemical Society. On August, 23rd, the ACS announced that Dr. Lipinski is the recipient of the 2005 E.B. Hershberg award for improved discoveries in medicinally active substances. Since its publication in 1997, the Lipinski's rule of five has been a critical filter for drug development programs. A simple algorithm that helps identify successful drug candidates, the principles filter out molecules likely to have poor intestinal permeability or poor aqueous solubility and hence poor oral absorption. This landmark contribution to drug development has influenced the way that the pharmaceutical industry approaches the development of orally active drugs. Drug discovery programs worldwide use the rule as a filter in high-throughput screening libraries and the T.B. Aliance is applying Dr. Lipinski's Rule to its go/no-go decision making process for its projects in the nitroimidazopyan and quinolone classes.

The rule of five analyses helped to raise awareness about properties and structural features that make more or less drug-like. The rule describes molecular properties important for drug's pharmacokinetics in the human body, including their absorption, distribution, metabolism and excretion ("ADME"). However, the rules do not predict whether a compound is pharmacologically active or not. The rule is important for drug development where a pharmacologically active lead structure is optimized step-wise for increased activity and selectivity as well as drug-like properties as described by Lipinski's rule. Modification of the molecular structure often leads to drugs with higher molecular weight, more rings, more rotatable bonds and a higher lipophilicity.

The Lipinski's Rule of Five **(Lipinski *et al.*, 1997)** states that an orally active drug should obey the following criteria:

1. Not more than 5 hydrogen bond donors
2. Not more than 10 hydrogen bond acceptors
3. A molecular weight under 500g/mol
4. A partition coefficient log P less than 5
5. Number of rotatable bonds not less than 10

Lipinski's work has since been extended to include properties such as the number of rings and rotatable bonds. Later on, several groups have researched and published profiling techniques to further define drug likeness **(Oprea *et al.*, 2001)** and lead likeness **(Arup *et al.*, 1999)** in order to increase the efficiency of drug discovery. The Rule of Five is so called because most of the features start with the number of five, close to 5 or a multiple of five.

About Dr. Lipinski

Dr. Lipinski joined Pfizer Company in 1970 supervising medicinal chemistry laboratories, discovering multiple gastrointestinal and diabetic clinical candidates. Since 1984 he has been an adjunct faculty member at Connecticut college in New London, Connecticut. During the course of his distinguished career he has authored over 190 publications and invited presentations and has 17 issued U.S. patents.

In 1990, he established a laboratory combining computations and experimental physical property and since 2001 has been a member of the scientific advisory committee of the T.B. Alliance, helping to guide the research and development f novel T.B. drugs. Dr. Lipinski obtained his Ph.D. from the University of California, Berkeley, and did his Postdoctoral training at Caltech, supported by the National Institute of General Medical Sciences.

Selected Drugs for the Present Study

There is a mounting need for therapeutics to effectively treat neurodegenerative diseases. Alzheimer's Disease, Parkinson's Disease, Huntington's Disease, amylotropic Sclerosis and Multiple Sclerosis almost all share pharmacological hallmarks of accumulated misfold protein, ultimately leading to cellular degeneration and death **(Pangalos *et al.*, 2007)**. There is much to be learned by the successes and failures of drug discovery efforts for these respective diseases. Exciting and novel ides from academia often fail to reach drug discovery platforms and pharmaceutical companies have had little success in their neurodegenerative disease programs. Thus, currently only symptomatic treatments are available for the majority of these diseases **(Ringman and Cummings, 2006; Scatena *et al.*, 2007)**.

While these diseases present unique challenges in terms of drug discovery, they also offer many opportunities to change the way academics and industry work together to efficiently develop new drugs. The conference held during February 2-3rd in Washington was hosted by the Alzheimer's Drug Discovery foundation and during this conference, speakers presented lectures and case studies on neurological diseases like Alzheimer's disease and shared common challenges, co-ordinated multidisciplinary approaches required for novel discoveries on effective therapeutics.

In the present study, a few drugs such as galantamine, donepezil, tacrine, ibuprofen and metriphonate which are currently in use to treat Alzheimer's disease were selected to test whether they follow the Lipinski's rule of five.

Galantamine

It was approved by the FDA in Feb, 2001 for the treatment of mild to moderate dementia of the Alzheimer's type. It is postulated to exert its effect by enhancing cholinergic function through a competitive and reversible inhibition of acetylcholinesterase in the central nervous system.

Donepezil

Donepezil marketed under the trade name Aricept by its developer Eisai and partner Pfizer, is a centrally acting reversible acetylcholinesterase inhibitor **(Birks and Harvey, 2006)**. Its main therapeutics use is in the treatment of Alzheimer's disease where it is used to increase cortical acetylcholine.

Tacrine

Tacrine is a parasympathomimetic and a centrally acting cholinesterase inhibitor. It was the first centrally-acting cholinesterase inhibitor approved for the treatment of Alzheimer's disease and was marketed under the trade name Cognex. Tacrine was first synthesized by Adrein Albert at the University of Sydney.

Ibuprofen

Ibuprofen was derived from Propionic acid by the research arm of Books group during the 1960's **(Adams, 1992)**. In some studies ibuprofen showed superior results compared to a placebo in the prophylaxis of Alzheimer's disease when given in low doses over a long time **(Townsend and Practico, 2005)**.

Metriphonate

Metriphonate or Trichloforn is an organophosphate acetylcholinesterase inhibitor. It is classified as irreversible (National Library for Health).

According to Lipinski, the drugs which follow these rules are available in market. In the present study we selected some Alzheimer's drugs which were designed before the Lipinski's rule of five was put forth and checked whether they satisfy the rule or not with the help of appropriate bioinformatics softwares.

Plan of work

The main aim of the present work is "validation of Lipinski rule of five for different drugs used to treat Alzheimer's Disease". In the present scenario simple compounds like Galantamine, Donepezil, Tacrine, Ibuprofen and Metriphonate are being used to treat Alzheimer's Disease. For the present work, validation of chemical drugs was done by:

a) Random selection of different drugs used to treat Alzheimer's Disease.

b) Obtaining SMILES format for these drugs using softwares.

c) Applying 'Lipinski's rule of five' for each of these drugs.

d) Finally, resolving violation of the Lipinski rule.

Softwares Used

Molinspiration

Molinspiration is an independent research organization focussing on development and application of modern cheminformatics techniques, especially in connection with the internet. Molinspiration offers broad range of cheminformatics software tools supporting molecule manipulation and processing, including SMILES and SDfile conversion, normalization of molecules, generation of tautomers, molecule fragmentation,calculation of various molecular properties needed in QSAR, molecular modelling and drug design, high quality molecular depiction, molecular database tools supporting substructure search or similarity and pharmacophore similarity search. Molinspiration tools are written in Java and therefore are available practically on any computer platform.

Molinspiration also supports internet chemistry community by offering free-on-line cheminformatics services for calculation of important molecular properties (for example logP, Polar surface area, number of hydrogen bond donors and acceptors) as well as prediction of bioavailability score for the most important drug targets.

The picture below shows all the components of Lipinski's rule i.e.

- miLogP - A partition coefficient log P
- MW - Molecular Weight
- nON - Not more than 10 hydrogen bond acceptors
- nOHNH - Not more than 5 hydrogen bond donors

It also gives a plan of:

- natom - number of heavy atoms
- TPSA - Total Polar Surface Area
- nrotb - number of rotatable bonds
- nviolations - number of violations made by the proposed drug (nvio-used later)
- volume - the volume of the compound (vol. – used later)

SMILESCOc1cc2C(=O)C(Cc2cc1OC)CC4CCN(Cc3ccccc3)CC4

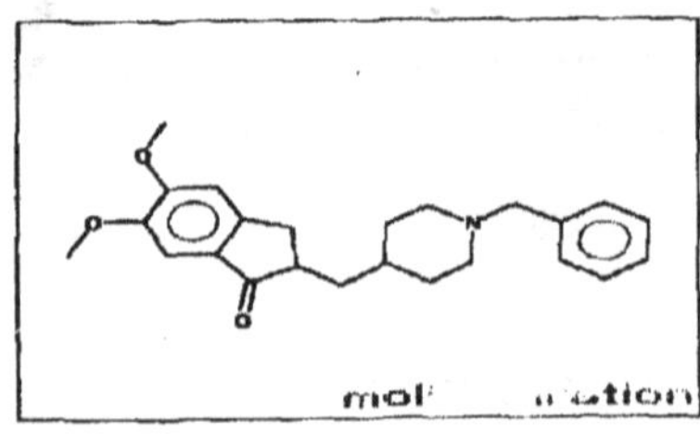

miLogP	4.1
TPSA	38.777
natoms	28
MW	379.5
nON	4
nOHNH	0
nviolations	0
nrotb	6
volume	367.895

The software also gives the canonical smiles of the same chemical being used.

LogP (Octano/water partition coefficient)

LogP actually describes a disolvation effect where the molecule in their hydration spheres need to undergo disolvation at the receptor to realize the hydrogen bonding and dipole interactions with the receptor, in general terms, the BBB exterior membrane. Log P is calculated by the methodology developed by Molinspiration as a sum of fragment-based contributions and correction factors. This method is very robust and is able to process practically all organic and most organometallic molecules.

Total Molecular Polar Surface Area TPSA

PSA is defined as the surface area ($Å^2$) occupied by nitrogen and oxygen atoms and the polar hydrogens attached to them and is strongly reflective of hydrogen bonding capacity and polarity. Both of these forces are involved as the molecule approaches the polar surface of the membrane and desolvates as it moves into the lipid portion. It is calculated based on the methodology published by Ertl *et al.*, 2000 as a sum of fragment contributions. O- and N- centered polar fragments are considered. PSA has been shown to be very good descriptor characterizing drug absorption including intestinal absorption, bioavailability, Caco-2 permeability and blood-brain barrier penetration.

Total Number of Atoms

The sum of the (N + O) atoms actually measures the hydrogen bond acceptors associated with nitrogen and oxygen moieties.

Molecular Volume

Molecular volume is a function of MW and structure and takes into account all the accessible conformations available to the molecule under physiological conditions. This actually relates to rotatable bonds and the number of rings in the molecule. Method for calculation of molecular volume, developed at Molinspiration is based on group contributions. These have been obtained by fitting sum of fragment contributions to "real" 3D volume for a training set of about twelve thousand, mostly drug-like molecules. 3D molecular geometrics for a training set were fully optimized by the semi empirical AM1 method.

"Rule of 5" properties

These are a set of simple molecular properties used by Lipinski in formulating his "Rule of 5" (Lipinski et al., 1997). The rule states that most "drug-like" molecules have logP < 5, molecular weight < 500, number of hydrogen bond acceptors < 10, and number of hydrogen bond donors < 5. Molecules violating more than one of these rules may have problems with bioavailability. The rule is called "Rule of 5" since the border values are not considered because of their high rotational energy barrier.

Results

Galantamine

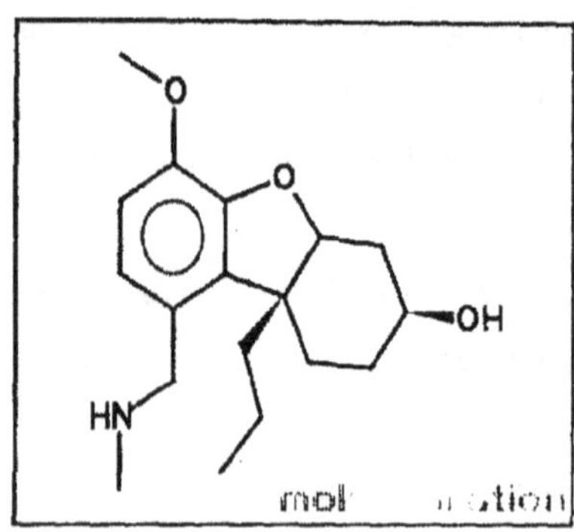

v2009.01

miLogP	2.623
TPSA	50.723
natoms	22
MW	305.418
nON	4
nOHNH	2
nviolations	0
nrotb	5
volume	301.402

SMILES CCC[C@]13CC[C@H](O)CC1Oc2c(OC)ccc(CNC)c23

Donepezil

v2007.04

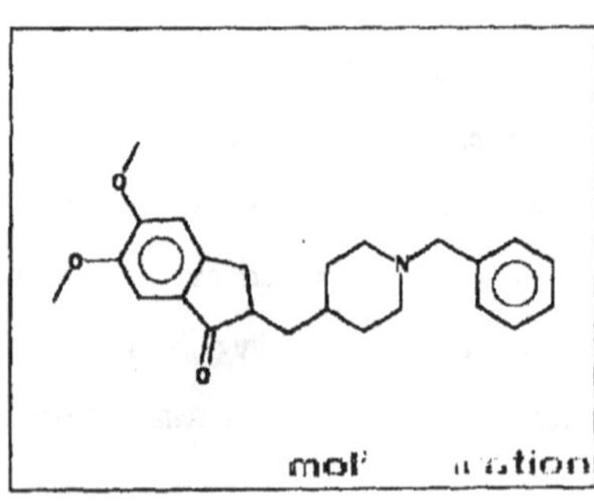

miLogP	4.1
TPSA	38.777
natoms	28
MW	379.5
nON	4
nOHNH	0
nviolations	0
nrotb	6
volume	367.895

SMILES COc1cc2C(=O)C(Cc2cc1OC)CC4CCN(Cc3ccccc3)CC4

Tacrine

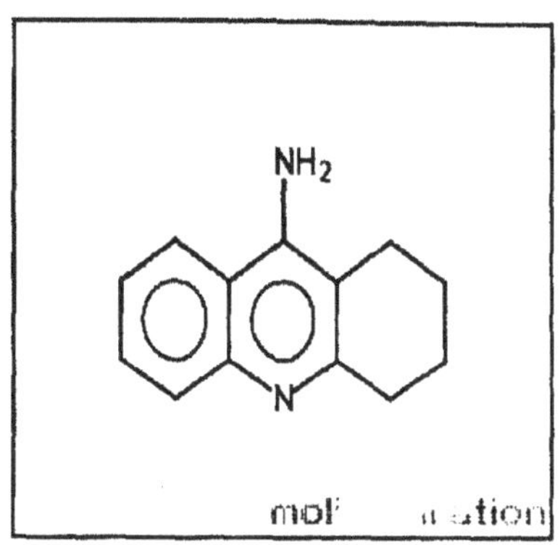

Molinspiration property engine
v2009.01

miLogP	3.05
TPSA	38.915
natoms	15
MW	198.269
nON	2
nOHNH	2
nviolations	0
nrotb	0
volume	191.533

SMILES Nc2c1CCCCc1nc3ccccc23

Ibuprofen

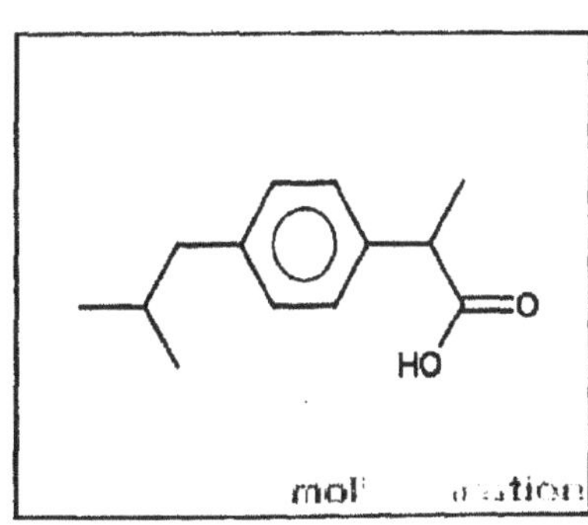

Molinspiration property engine
v2007.04

miLogP	3.462
TPSA	37.299
natoms	15
MW	206.285
nON	2
nOHNH	1
nviolations	0
nrotb	4
volume	211.185

SMILES CC(C(=O)O)c1ccc(CC(C)C)cc1

Metrifonate

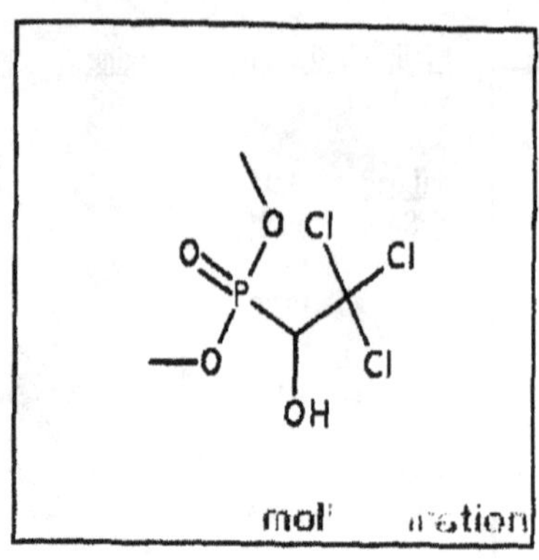

Molinspiration property engine v2009.01

miLogP 0.798
TPSA 55.767
natoms 12
MW 257.437
nON 4
nOHNH 1
nviolations 0
nrotb 4
volume 174.03

SMILES COP(=O)(OC)C(O)C(Cl)(Cl)

Table 12: Showing the properties of a few selected drugs obtained through molinspiration for the validation of the Lipinski Rule of 5

Name of the Drug	Chemical Formula	Milo g P	TPSA	natom	MW	nON	nNHOH	nvio	nrotb	Vol.
Galantamine	$C_{17}H_{21}NO_3$	2.6	50.7	22	305.4	4	2	0	5	301.402
Donepezil	$C_{24}H_{29}NO_3$	4.1	38.7	28	379.5	4	0	0	6	367.895
Tacrine	$C_{13}H_{14}N_9$	3.0	38.9	15	198.2	2	2	0	0	191.533
Ibuprofen	$C_{13}H_{18}O_2$	3.4	37.2	15	206.2	2	1	0	4	211.185
Metriphonate	$C_4H_8Cl_3O_4P$	0.7	55.7	12	257.4	4	1	0	4	174.03

- From the above table, it was observed that among the five drugs selected, Donepezil has high milogP value (4.1) followed by Ibuprofen (3.462), Galantamine (2.623), Tacrine (3.05) and Metriphonate (0.798).

The Total Polar Surface Area was recorded highest in Metriphonate (55.567) and Galantamine (50.723), Tacrine (38.915), Donepezil (38.777) and Ibuprofen (37.299) had medium TPSA.

More number of atoms are presented in Donepezil (28) followed by Galantamine (22), Tacrine, Ibuprofen (15) and Metriphonate (12). Further, it is obvious that Donepezil has highest molecular weight (379.5) followed by Galantamine (305.418), Metriphonate (257.437), Ibuprofen (206.285) and Tacrine (198.269).

While Galantamine, Donepezil and Metriphonate have 4 nON's, Tacrine, Ibuprofen have only 2. Galantamine, Tacrine have 2 nNHOH, Ibuprofen, Metriphonate have 1 nNHOH and Donepezil does not have any nNHOH.

The highest rotatable bonds are present in Donepezil (6) followed by Galantamine (5) and Ibuprofen and least in Metriphonate (4).

Apart from these, it was also observed that Donepezil has high volume (367.895) followed by Galantamine (301.402), Ibuprofen (211.185), Tacrine (191.533) and Metriphonate has least (174.03).

Discussion:

In the broadest sense, moderately lipophilic drugs cross the Blood Brain Barrier by passive diffusion and the hydrogen bonding properties of drugs can significantly influence their Central Nervous System uptake profiles. Polar molecules are generally poor Central Nervous System agents unless they undergo active transport across the Central Nervous System. Size, ionization properties and molecular flexibility are other factors observed to influence the transport of an organic compound across the Blood Brain Barrier **(Mouristen and Jorgensen, 1998).**

An orally active anti-Alzheimer's drug needs not only sufficient metabolic stability to maintain integrity in the intestine and liver but also across the BBB. At the molecular level, the BBB is not homogenous but consists of a number of partially overlapping zones contained in a highly anisotropic lipid layer **(Sippl, 2002).** The conformational mobility of the lipid chains is relatively low at or near the water (blood)/ lipid interface and interface at center of the bilayer. The lipid – water interface is associated with a layer of perturbed water molecules with significantly different polarization properties. Because of this, the ability of these water molecules to form hydrogen bonds with drug molecules is dramatically reduced and forms part of the disolvation process. In addition, the hydrophilic/lipophilic interface at the blood/membrane boundary consists of perturbed and bound water, charged polar lipid head moieties connected to long lipid chains. As a result, a drug approaching the BBB is confronted with a thick layer that is capable of noncovalent interactions with the drug, similarly to that of receptor but with much looser steric requirements.

Optimizing the chemical structure of lead candidates with respect to the ADME processes has become an integral part of the drug discovery paradigm **(Dennis, 1990).** An important ADME characteristic is simply the solubility of the drug, as only the amount of drug in solution is available for intestinal absorption and blood distribution **(Kerns and Di, 2003).** The initial analyses of ADME properties, e.g. anesthetic agents in the late nineteenth century, focused on the partition coefficient (LogP) between water and oil, basically the lipophilicity of the compound. This has served as one of the fundamental principles for drug discovery and design **(Gupta, 1989).** High lipophilicity frequently leads to compounds with high rapid metabolic turnover **(Van de waterbeemed et al., 2001)** and low solubility and poor absorption. As lipophilicity (LogP) increases, there is an increased probability of binding to hydrophobic protein targets other than the desired one, and therefore; there is more potential for toxicity.

Hansch's initial work was based on a large set of sedative-hypnotic barbituates where he demonstrated biological activity was almost entirely due to their Log P and their rate of metabolism was linearly related to LogP. Furthermore, optimal activity is observed at LogP = 2 **(Hansch et al., 1967)**. The drugs used to treat neurological disorders have LogP value mostly between 2 to 4 **(Young et al., 1986)**. Subsequently, Goodwin et al., indicated that LogP is predominantly a measure of drug volume or surface area, plus hydrogen bond acceptor potential. Thus, both hydrogen bonding potential and drug volume contribute to permeability. Mostly CNS drugs have these properties.

Lipophilicity was the first of the descriptors to be identified as important for CNS penetration. Hansch and Leo reasoned that highly lipophilic molecules will be partitioned into the lipid interior of membranes and will be retained there. However, ClogP correlates nicely with LogBBB with increasing lipophilicity increasing brain penetration. For several classes of CNS active substances, Hansch and Leo (1987) found that blood-brain barrier penetration is optimal when the LogP values are in the range of 1.5-2.7, with the mean value of 2.1. An analysis of small drug-like molecules suggested that for better brain permeation **(Van de waterbeemed et al., 1998)** and for good intestinal permeability **(Fischert et al., 2003)** the LogP values need to be greater than 0 and less than 3.

The Polar Surface Area (PSA) and the molecular volume components were the most important descriptors, with PSA strongly predominating **(Van de Waterbeemd and Kancy, 1992)**. Palm *et al.*, 1999 developed a dynamic PSA approach whereby the set of available conformations were used and the contribution of each to the overall PSA was calculated using a **Boltzman** distribution thereby taking into account conformational flexibility. Based on their results in intestinal Caco-2 cells, drugs with a PSA of 60 $Å^2$ or less are completely absorbed, whereas those with at least 140 $Å^2$ are not. PSA calculation gave good agreement and separation based on PSA between the CNS and non-CNS drugs. Kelder (1999) found that non-CNS drugs transported passively and transcellularly needed a PSA of 120 $Å^2$ or less, whereas the drugs can be targeted to the CNS with a PSA less than 60–70 $Å^2$. Similar conclusions were made by van de Waterbeemd based on a study of marketed CNS and non-CNS drugs **(Van de waterbeemd et al., 1998).** Their cutoff for PSA cutoff for CNS penetration is 90 Å^2 or below and a molecular weight cutoff of 450.

Ertl has developed a topological PSA (TPSA) approach that fits these criteria **(Ertl et al., 2000).** TPSA is based on dissecting the contributions of polar groups in drugs contained in the WDI. Comparison with Clark's results demonstrated almost no difference between the two approaches. The major advantage of TPSA is that it uses a two-dimensional structure and not the computationally more intensive three-dimensional representation. This enables two to three orders of magnitude increases in throughput.

Analysis has also shown that a large PSA **(Lipinski et al., 1997; Clark et al., 2000)** (greater than 150-200 Å^2) or rotatable bonds **(Veber et al., 2002)** beyond 10 lead to dramatically decreased permeability and oral bioavailability.

Because PSA is dependent upon hydrogen bonding and donating atoms, Österberg **(Österberg and Norinder, 2000)** investigated simplifying the PSA calculation by incorporating hydrogen bonding descriptors in place of PSA while including ClogP. There was high correlation between the two methods, indicating the importance of hydrogen bonding and lipophilicity for LogBBB. As with TPSA, this simplification allows BBB penetration evaluation of large actual and virtual libraries.

Hydrogen bonding approach was later extended and developed into a pair of rather simple rules for predicting BBB penetration:

Because hydrogen bonding is primarily associated with oxygen and nitrogen moieties in a molecule, then, if the sum of the nitrogen (N) and oxygen (O) atoms in the molecule is five or less, then the molecule has a high probability of entering the CNS.

$$\text{BBB penetration} = (N+O) = \leq 5$$

PSA has been used as a predictor for BBB penetration by many investigators **(Lenz, 1999; Feng, 2002).** In general, drugs aimed at the CNS tend to have lower polar surface areas than other classes **(Skaaeda et al., 2001; Abraham et al., 1999).** PSA for CNS drugs is significantly less than for other therapeutics with PSA for CNS penetration estimated at 60–70 Å^2 through 90 Å^2 **(Van de Waterbeemd et al., 1998).** The upper limit for PSA for a molecule to penetrate the brain is around 90 Å^2.

In an interesting study that proved prescient, Levin determined that rat brain permeability was determined by LogP and had a molecular weight cutoff of 400 or less **(Levin, 1980).** CNS drugs have significantly reduced molecular weights (MW) compared with other therapeutics. Levin has a cutoff for BBB penetration of 400. Van de Waterbeemd (2002) has suggested that MW should be kept below 450 to facilitate brain penetration and to be lower than that for oral absorption **(Van de Waterbeemd, 1998).** The rules for molecular weight have been reviewed, where small molecules may undergo significant passive lipid-mediated transport through the blood brain barrier, when the molecular mass is kept in or below a 400- to 600-Da range **(Hansch et al., 1967).** For marketed CNS drugs, the mean value of MW is 310, compared with a mean MW of 377 for all marketed orally active drug **(Leeson and Davis, 2004).**

All the QSAR equations emphasize the importance of hydrogen bonding whether through polarity, PSA, hydrogen bond donor and acceptor counts, or simply counting heteroatoms capable of hydrogen bonding. All of these measurements are correlated, for instance, (O + N) atom count is highly correlated with PSA but measures hydrogen bond acceptors. CNS penetration requires a sum of these heteroatoms of 5 or less **(Österberg and Norinder, 2000)** Compounds with high hydrogen bond forming potential, such as peptides with their amide groups, peptides even as small as di- or tripeptides, have minimal distribution through the BBB **(Pardridge, 1998).** Increasing hydrogen bonding decreases BBB penetration. The marketed CNS drugs on average have (O + N) = 4.32 **(Leeson and Davis, 2004).** It should be pointed out that there are other heteroatoms in drugs that can function as hydrogen bond acceptors (HBA) and total HBA, including (N + O) would probably give a better measure.

Rotatable bond count is now a widely used filter following the finding that greater than ten rotatable bonds correlates with decreased rat oral bioavailability **(Veber et al., 2002).** CNS drugs have significantly fewer rotatable bonds than other drug classes. Most centrally acting compounds have rotatable bond count of five or less **(Leeson and Davis, 2004).**

Lipinski's rule of five is a heuristic approach for predicting drug - likeness stating that molecules having M.W. > 500, LogP > 5, hydrogen bond donors > 5 and hydrogen bond acceptors > 10 have poor absorption or permeation **(Lipinski, 2000).**

Everyday like these so many pharmacological companies concentrate on the synthesis of new drugs to treat Alzheimer's Disease. Some drugs are available in market for many years some are eliminated in very few days. Because they are having different properties like high molecular weight and number of hydrogen atoms donors and acceptors. Those drugs violated like the "Rule of Five". For these reasons in the present study we have selected some Alzheimer's drugs and test whether they follow the Lipinski's rule of five and test the results of the drug validation. For this selected five drugs were taken from DRUGBANK. The drugs are Galantamine, Donepezil, Tacrine, Ibuprofen and Metriphonate. These drugs are validated according to the Lipinski's rule of five. A table was prepared for easy reference.

It was interesting that none of the drugs violated the Lipinski's rule of five. It was also a matter of concern that though some of the withdrawn drugs had no violations they were removed from the market. The majority of drugs are intended for oral therapy. Therefore obtaining oral activity may be the rate limiting step.

General Conclusion

The present study demonstrates the adverse effects of galantamine hydrobromide on morphometric and behavioural aspects, aminergic system and energy metabolism of mice on prolonged treatment for one year. The morphometric and behavioural changes in treated mice showed a positive result upto around 240 days and then registered a gradual loss in their body weights and declined activity levels. These behavioural changes were very well coincided with the perturbed trend observed in the aminergic system and also the energy metabolism the details of which were presented in the chapter II and III.

One of the most significant observation in the present study was that treatment of mice with Galantamine hydrobromide infact improved the learning capabilities of mice from one month up to six months (corresponding age of human 2.5 years to 15 years as given in the table in materials and methods section). Contrary to this, between six to twelve months (corresponding age of human between 16 years to 30 years), this drug induced adverse effects on the overall performance of the mice. This observation has direct relevance to human beings. It has been reported that the children between 1 to 12 of age have highest perception of learning and memory. Having known this fact, parents give memory boosters, health drinks or drugs containing such similar substances available in the market which might improve the overall performance of the children in their early stages of career but after 25th or 30th year onwards the chemicals in these drugs or health drinks start exerting ill effects on all organ systems in general and the nervous system in particular. In view of these observations, it is not advisable to recommend these health drinks or drugs as memory boosters particularly to children.

CPSIA information can be obtained
at www.ICGtesting.com
Printed in the USA
BVHW031407270423
663156BV00007B/357